Essentials of Food Safety and Sanitation

Food Safety Fundamentals

2nd Edition

David McSwane, H.S.D.

Richard Linton, Ph.D.

Nancy R. Rue, Ph.D.

Anna Graf Williams, Ph.D.

Production Services and Development by

Learnovation®, LLC

FOOD MARKETING INSTITUTE

Editor-in-Chief: Lawrence R. Kohl

Editorial Assistant: Gwendolyn Lee

Product Liaison: Dr. Jill Hollingsworth

Production Editor: Anna Graf Williams

Senior Design Coordinator: Karen J. Hall

Copy Editor: Cheryl Pontius

Cover Design: John A. Lezcano

10 9 8 7 6 5 4 3 2 1

ISBN: 978-0-9819903-7-8

FOOD MARKETING INSTITUTE

$79.99

ISBN 978-0-9819903-7-8

57999>

9 780981 990378

Contents

Preface xiii

Acknowledgments xv

Chapter 1– Food Safety and Sanitation Management   2

New Challenges Present New Opportunities 4

Food Safety–Why All the Fuss? 4

Why Me? 5

Changing Trends in Food Consumption and Choices 6

The Problem: Foodborne Illness 7

    Contamination 8

    Microorganisms 9

The Food Flow 9

Facility Planning and Design 10

Keeping It Clean and Sanitary 10

A New Approach to an Old Problem 10

Accident Prevention and Crisis Management 11

Education and Training Are Key to Food Safety 11

The Role of Government in Food Safety 12

    The FDA *Food Code* 12

The Role of the Food Industry in Food Safety 12

Food Protection Manager Certification 12

Summary 14

Quiz 1 (Multiple Choice) 15

References/Suggested Readings 16

Suggested Web Sites 16

Chapter 2– Hazards to Food Safety   18

Foodborne Illness 20

Foodborne Hazards 22

    Bacteria 22

    Spoilage and Disease-causing Bacteria 23

    Bacterial Growth 23

What Disease-causing Bacteria Need in Order to Multiply 24

    Food 25

    Acidity 25

    Temperature 25

    Time 26

Oxygen 26

Moisture 26

Potentially Hazardous Food (Time/Temperature Control for Safety Food) 28

Ready-to-Eat Foods 30

Foodborne Illness Caused by Bacteria 30

Foodborne Illness Caused by Sporeforming Bacteria 31

Foodborne Illness Caused by Non-sporeforming Bacteria 36

Foodborne Illness Caused by Viruses 43

Foodborne Illness Caused by Parasites 45

Problems Caused by Fungi 48

Foodborne Illness Caused by Chemicals 49

Naturally Occurring Chemicals 49

Food Allergens 49

Added Man-Made Chemicals 53

Foodborne Illness Caused by Physical Hazards 54

International Food Safety Challenges 55

Summary 55

Quiz 2 (Multiple Choice) 56

References/Suggested Readings 58

Suggested Web Sites 58

Chapter 3– Factors That Affect Foodborne Illness   60

Agents that Lead to Foodborne Illness in the United States 62

Factors That Contribute to Foodborne Illness 63

What Is Temperature and Time Abuse? 64

Measuring Food Temperatures Correctly 65

When and How to Calibrate a Thermometer 66

Ice Point Method 67

Boiling Point Method 67

Measuring Food Temperature 67

How to Accurately and Safely Measure Food Temperatures 67

Preventing Temperature and Time Abuse 68

Keep Cold Foods Cold and Hot Foods Hot! 71

Proper Thawing of Food 72

The Importance of Good Personal Hygiene and Hand Washing 73

Hand Washing 73

Hand Antiseptics (Hand Sanitizers) 75

Using Disposable Gloves 75

Outer Clothing and Apparel 75

Personal Habits 76

Personal Health 77

Cross Contamination 78

Preventing Cross Contamination 78

Other Sources of Contamination 79

Make Sure the Work Area Is Clean and Sanitary 79

Summary 80

Quiz 3 (Multiple Choice) 81

References/Suggested Readings 83

Suggested Web Sites 83

Chapter 4– Following the Flow of Food   86

Following the Flow of Food 88

Buying from Approved Sources 89

Measuring Temperatures During Receiving, Storage and Preparation 89

Receiving 89

Inspecting Delivery Vehicles 89

Determining Food Quality 90

Packaged Foods 91

Reduced Oxygen Packaging (ROP) 92

Irradiation 95

Red Meat Products 95

Poultry 96

Game Animals 97

Eggs 98

Fluid and Dry Milk and Milk Products 99

Fluid Milk 100
Cheese 100
Butter 100

Fish 100

Fruits and Vegetables 103

Juice 104

Frozen Foods 104

Storage of Food 104

Types of Storage 105

Refrigerated Storage 105
Freezer Storage 106
Dry Storage 107
Chemical Storage 107

Storage Conditions for Foods 108

Preparation and Service 110

Ingredient Substitution 110

Avoiding Temperature Abuse 110

Thawing 110

Cold Storage 111

Frozen, Ready-to-Eat Foods 113

Prepackaged Foods 113

Cooking 113

Cooling 115

Hot-holding and Reheating 117

Time as a Public Health Control 117

Serving Safe Food 118

Discarding or Reconditioning Food 118

Refilling Returnable Containers 118

Self-Service Bar 119

Rules for Self-Service Bars: 119

Temporary Facilities and Mobile Food Facilities 119

Vending Machines 120

Home Meal Replacement 121

Summary 122

Quiz 4 (Multiple Choice) 123

References/Suggested Readings 125

Suggested Web Sites 125

Chapter 5– Facilities, Equipment, and Utensils   128

Design, Layout, and Facilities 130

Regulatory Considerations 130

Work Center Planning 131

Equipment Selection 131

Size and Design 131

Construction Materials 132

Types of Equipment 134

Cooking Equipment 135

Ovens and Fryers 135

Refrigeration and Low Temperature Storage Equipment 136

Reach-In Refrigeration 138

Cook-chill and Rapid-chill Systems 139

Hot-Holding Equipment 139

Other Types of Food Equipment 140

Slicers 140

Mixers, grinders, and choppers 140

Ice machines 141

Single-Service and Single-Use Articles 142

Warewashing Equipment 142

Manual Warewashing 143

Mechanical Warewashing 143

Installation 144

Maintenance and Replacement 144

Lighting 144

Heating, Ventilation, and Air Conditioning (HVAC) 146

Summary 147

Quiz 5 (Multiple Choice) 148

References/Suggested Readings 150

Suggested Web Sites 150

Chapter 6– Cleaning and Sanitizing Operations   152

Principles of Cleaning and Sanitizing 154

Cleaning Principles 155

Pre-scrape / Pre-flush to Remove Food Particles 155

Application of Cleaning Agents 155

Soaking 156
Spray Methods 156
Clean-in-place systems 156
Abrasive cleaning 156

Detergents and Cleaners to Be Used 156
Factors that affect cleaning efficiency 156

Rinsing 157
Cleaning frequency 157

Sanitizing Principles 159

Heat Sanitizing 159
Hot Water 160
Steam 160

Chemical Sanitizing 161

Mechanical Warewashing 163
Mechanical warewashing process 163

Manual Warewashing 165

Cleaning Fixed Equipment 166
Wiping cloths 168

Cleaning Environmental Areas 169

Ceilings and Walls 169

Floors 169

Equipment and Supplies Used for Cleaning 170

Summary 172

Quiz 6 (Multiple Choice) 173

References/Suggested Readings 174

Suggested Web Sites 175

Chapter 7– Environmental Sanitation and Maintenance 178

Condition of the Food Establishment 180

Proper Water Supply and Sewage Disposal System 180

Condition of Building 181

Floors, Walls, and Ceilings 182

Floors 182
Walls and Ceilings 183

Restroom Sanitation 183

Handwashing Facilities 184

Plumbing Hazards in Food Establishments 185

Cross Connections and Backflow 185

Methods and Devices to Prevent Backflow 185

Backflow Prevention Devices on Carbonators 187

Grease Traps 187

Garbage and Refuse Sanitation 187

Inside Storage 188

Outside Storage 188

Pest Control 189

Integrated Pest Management (IPM) 189

Insects 189

Rodents 191

Signs of Rodent Infestation 191
Rodent Control 193

Summary 194

Quiz 7 (Multiple Choice) 195

Suggested Reading/References 197

Suggested Web Sites 197

Chapter 8– Food Safety Management Programs and the HACCP System 200

The Problem 202

The Solution 202

The Hazard Analysis Critical Control Point (HACCP) System 203

Product Description and Development of the Food Flow Diagram 203

The Seven Principles in a HACCP System 206

Principle 1—Hazard Analysis 206

Principle 2—Identify Critical Control Points (CCPs) 208

Principle 3—Establish the Critical Limits Which Must Be Met at Each Critical Control Point 210

Principle 4—Establish Procedures to Monitor CCPs 212

Principle 5—Establish the Corrective Action to Be Taken When Monitoring Shows a Critical Limit Has Been Exceeded 213

Principle 6—Establish Procedures to Verify the HACCP System Is Working 213

Principle 7—Establish an Effective Record-Keeping System That Documents the HACCP System 214

Roles and Responsibilities Under HACCP 215

Complementary Food Safety Management Programs 215

Coding and Product Identification 216

Food Recalls 216

Food Safety Related Crisis Management Situations 217

Water Supply Emergency Procedures 217

Foodborne Illness Incident or Outbreak 218

Bioterrorism and Food Protection 219

Challenges of Food Safety Management 220

Summary 221

Quiz 8 (Multiple Choice) 222

References/Suggested Readings 224

Suggested Web Sites 224

Chapter 9– Food Safety Regulations 226

State and Local Regulations 228

Permit to Operate 228

Routine Inspections 229

Inspection Provisions 230

Federal Agencies 231

Other Food Safety Related Organizations 232

Inspection for Wholesomeness and Grading of Food 233

Food Labeling 233

Safe Food-Handling Label 234

Food Allergen Labeling and Consumer Protection Act 234

Product Dating Labels 235

Date Marking for Ready-to-Eat, PHF (TCS) 235

Consumer Advisory 237

Information about Food Safety Guidelines and Regulations 238

The Fight BAC!™ Campaign 238

Be Food Safe 238

Additional Information Web sites 239

Summary 240

Quiz 9 (Multiple Choice) 240

References/Suggested Reading 242

Web Sites 242

Glossary 245

Appendixes 261

Appendix A 261

Answers to End of Chapter Questions 261

Appendix B 263

The Process for Determining Potentially Hazardous Food (Time/Temperature Control for Safety Food [PHF (TCS)]) 263

Appendix C 267

Summary of Agents That Cause Foodborne Illness 267

Appendix D 271

Multitiered, Risk-based Employee Health System 271

2-201.11 Responsibility of Permit Holder, Person in Charge, and Conditional Employees 272

2-201.12 Exclusions and Restrictions 274

2-201.13 Removal, Adjustment, or Retention of Exclusions and Restrictions 275

Appendix E 281

Conversion Table for Fahrenheit and Celsius for Common Temperatures Used in Food Establishments 281

Appendix F 283

Areas of Knowledge Deemed Important for the Person in Charge 283

Appendix G 285

New Risk Designations for Food Code Provisions 285

Appendix H 287

Factors Affecting Cleaning and Sanitizing 287

Index 291

Take Note 301

Preface

Food safety and sanitation are very important issues for the success of the food industry. Customers expect and deserve to be served or sold safe and wholesome food, and a foodborne disease outbreak can not only harm the customers, it can also ruin your business.

Food can be contaminated at several points along the flow of food from production to consumption. It can be contaminated where it was produced or at food processing plants. It can also be contaminated while being transported and during final preparation at food establishments. Foods can also become contaminated by consumers in their homes.

Food establishments provide a last line of defense in controlling or eliminating the hazards that cause foodborne illness. Food establishments must handle foods and food ingredients safely and prepare food in a manner that reduces the risk of contaminated food being served or sold to your customers.

Effective food safety programs require managers, supervisors, and people in charge to have knowledge about the hazards associated with contaminated food. The person in charge must be committed to implementing safe food-handling practices in their establishment. Assuring food safety also requires trained employees who understand proper hygiene and food-handling practices and who will not take shortcuts when it comes to food safety.

The authors have created a text that provides "need to know" food safety information for food managers. **This book is based on the U.S. Food and Drug Administration's *2009 Food Code*.** It has been created to meet the training needs of restaurants, catering companies, health-care facilities, schools, corrections facilities, vending companies, convenience stores, supermarkets, other types of food establishments, and students seeking food safety education. We recommend *Food Safety Fundamentals* and our supplemental training materials for all of the following training activities:

> **The authors have created a text that provides "need to know" food safety information for food managers.**

◆ Short courses for food establishment managers in preparation for a national food protection manager certification examination

◆ Food safety and sanitation courses in vocational and culinary arts programs

◆ Self-study programs for food establishment managers who are preparing to take a national food protection manager certification examination.

One of the most important tasks you face as the person in charge is to train and supervise food employees. Your knowledge of food safety is of little value if you do not teach employees the correct way to handle food. You must always be on the lookout for situations where approved food-handling practices are not being followed. Violation of approved food safety guidelines and practices can endanger your customers and cause loss of reputation and financial harm to your establishment.

The team of McSwane, Linton, Rue, and Williams have produced a text to accompany the FDA's 2009 *Food Code*. We trust that you will find this text accurate, comprehensive, and, most of all, useful.

The authors of this textbook have been training food establishment managers and employees for over 30 years.

David Z. McSwane, H.S.D., REHS, CP-FS, is a nationally recognized trainer in food safety and sanitation. He has taught courses at the university level and for regulatory agencies, food establishments, food industry trade associations, vocational schools, and environmental health associations throughout the United States. Dr. McSwane is a recipient of the Walter S. Mangold award, the highest honor bestowed by the National Environmental Health Association. He is a member of the Environmental and Public Health Council at Underwriters Laboratories, Inc.

Richard Linton, Ph.D., is Professor of Food Safety and Director of the Center for Food Safety Engineering in the Food Science Department at Purdue University. He has provided research, education, and outreach leadership to the food industry for over 15 years. Dr. Linton is actively involved in developing and teaching food safety programs for the food industry including Hazard Analysis Critical Control Point (HACCP) programs, good manufacturing practices, good agricultural practices, the better process control school, sanitation, and safe food-handling practices for retail food establishments. His industry programs have been delivered throughout the United States and in many international countries.

Nancy Roberts Rue, Ph.D., R.N., has a background in teaching in technical education. Her doctorate is in educational leadership and curriculum and instruction from the University of Florida. With 30 years of experience in higher education and evaluation at Indiana University and St. Petersburg Junior College, she is dedicated to the task of building educational materials that meet the needs of those who want to learn. She is now an independent writer and consultant on training and development techniques.

Learnovation®, LLC, and Senior Partner **Anna Graf Williams, Ph.D.**, have provided expert instructional design and developmental editing for *Food Safety Fundamentals* as well as many other food safety texts over the last 10 years. Recognized for their team of experts with years of experience in instructional design, they have redesigned this second edition to include more tables, charts, call out boxes and photos to emphasize key points and make learning easier.

The authors wish all readers success in their food safety and sanitation activities. Regardless of where you work, always remember—foodborne illness is preventable. Follow the basic rules of food safety and you will enjoy a satisfying career in the food industry.

Acknowledgments

The authors wish to thank their families and colleagues who, through their support and patience, helped us with the completion of this project.

A greatful thanks goes to the FMI staff and FMI members who contributed to this project. Their collective technical expertise and insight into foodservice and supermarket operations and food safety programs make this manual second to none.

We would also like to thank White Lodging Services, Corp., Marsh Supermarkets, Inc., Crispers, LLC–Fresh Salads and Such, and Publix Super Markets, Inc., for their assistance in providing us with great locations and subjects for our photo shoots. Their photos have made it easier to demonstrate processes and key concepts throughout this new edition of the book.

Finally, the authors wish to thank all those persons who read this manual and implement its recommendations, making food safer every day.

Food Safety and Sanitation Management

Key Terms

- Clean
- Contamination
- FDA *Food Code*
- Food establishment
- Foodborne disease outbreak
- Foodborne illness
- Microorganisms (germs and microbes)
- Sanitary

You can protect the health and safety of your customers by developing and implementing effective food safety and sanitation practices within your food establishment. Learning about how food is contaminated and what actions are needed to prevent, control, and eliminate the agents that frequently cause foodborne illness and spoilage is the key to providing safe food.

Learning Points:

◆ Recognize how food safety and sanitation practices prevent foodborne illness in food establishments.

◆ State the problems caused by foodborne illness for individuals who become ill and the food establishment blamed for the incident.

◆ Identify trends in menus and consumer use of food products prepared in food establishments.

◆ Describe the role of government (federal, state, local, and tribal) in food safety.

◆ List the types of food establishments identified in this text.

◆ Describe the influence the FDA *Food Code* and state or jurisdictional codes can have on food establishments.

◆ Define the term Hazard Analysis Critical Control Point (HACCP) as applied in food safety management.

◆ Recognize the importance of food protection manager certification.

Salmonella outbreak linked to Vernon Hills Restaurant

Health officials say at least 163 people have been confirmed by laboratory analysis to be suffering from a *Salmonella* infection. Among the victims were 27 employees of the restaurant that is believed to be the source of the outbreak.

Investigators believe the outbreak was caused by an infected food employee at the restaurant who failed to wash his hands after using the bathroom. The ill employee then spread the *Salmonella* bacteria by touching food items and equipment with contaminated hands.

How could this outbreak have been prevented? See the summary at the end of this chapter.

New Challenges Present New Opportunities

The food industry is one of America's largest enterprises. It employs about one-fourth of the nation's work force and produces 20% of America's Gross Domestic Product (GDP). Billions of dollars worth of food are sold each year. Americans have made food a prominent part of their business and recreational activities.

The food industry is made up of businesses that produce, manufacture, transport, and distribute food for people in the United States and throughout the world. Food production involves many activities that occur on farms and ranches, in orchards, and in fishing operations. Food manufacturing takes the raw materials harvested by producers and converts them into forms suitable for distribution and sale. The distribution system consists of the many food operations that store, prepare, package, serve, display, vend, or otherwise provide food for human consumption. The term **food establishment** refers to all facilities involved in the food distribution chain, such as a restaurant, food market, institutional feeding location, or vending location

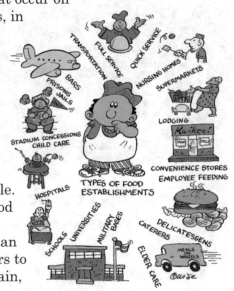

Food Safety–Why All the Fuss?

Everyone knows the United States has one of the safest food supplies in the world. So why all the fuss? The answer is simple. Foodborne illness happens, and it adversely affects the health of millions of Americans every year. Foodborne illness is the sickness some people experience when they eat contaminated food. It impairs performance and causes discomfort. Estimates of the number of cases of foodborne illnesses vary greatly. A report issued by the Centers for Disease Control and Prevention (CDC) estimates foodborne diseases cause approximately

DID YOU KNOW?

Foodborne illness costs billions of dollars each year in the form of:

◆ Medical expenses
◆ Lost work and reduced productivity by victims of the illness
◆ Legal fees
◆ Punitive damages
◆ Increased insurance premiums
◆ Lost business
◆ Loss of reputation for the food establishment.

76 million illnesses, 325,000 hospitalizations, and 5,000 deaths in the United States each year (Mead et. al., 1999).

THERE'S NEVER BEEN A CASE OF FOODBORNE ILLNESS THAT COULDN'T HAVE BEEN PREVENTED.

This book will help you look at food establishments and:

◆ Identify the risks to food safety
◆ Determine the actions to take to provide safe food
◆ Measure and monitor operations to be sure the necessary steps are taken to prevent foodborne illness.

Whether a person is an hourly employee or a member of the management team, their goal is the same: to provide safe food and prevent foodborne illness.

Prevention of foodborne illness must be a goal in every food establishment.

Check out the *Take Note* section starting on page 301. We've created a study guide to help you:

◆ Identify the risks;
◆ Take the right action, and
◆ Monitor tasks encountered in a food establishment.

We recommend you pull out the guide and use it as you read through the chapters. Take notes, fill in the information, answer questions, and identify the key information you need to know to prevent foodborne illness.

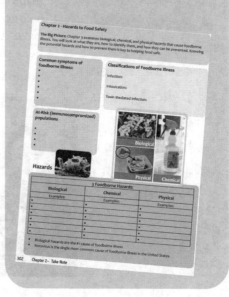

Why Me?

You may be asking yourself, "What does all this have to do with me?" The answer is "PLENTY." Customer opinion surveys show cleanliness and food quality are among the top reasons people use when choosing a place to eat and shop for food. Customers expect their food to taste good and be safe to eat. It is the responsibility of every food establishment owner, manager, and employee to prepare and serve safe and wholesome food and preserve their clients' confidence. As the person in charge, you must understand that foodborne illness can be prevented if the basic rules of food safety are routinely followed.

More than half of all foodborne disease outbreaks reported to the CDC are associated with food establishments. These outbreaks frequently result from foods being mishandled in retail food operations (restaurants, supermarkets, convenience stores, schools, churches, camps, health-care facilities, correctional facilities, and vending locations) where foods are prepared, served, and sold to the public.

These foods may be eaten at the food establishment or sold for preparation and consumption elsewhere.

notes

TOP 5 FACTORS CONTRIBUTING TO FOODBORNE ILLNESS:

■ **Improper holding temperatures (cooling, cold-holding, and hot-holding)**

■ **Inadequate cooking**

■ **Poor personal hygiene of food workers**

■ **Contaminated equipment**

■ **Food obtained from unsafe sources**

Source: FDA 2009 *Food Code*

Changing Trends in Food Consumption and Choices

More than 50% of the food eaten by Americans is prepared in:

◆ **Food-processing plants**

◆ **Restaurants**

◆ **Retail food establishments**

◆ **Delicatessens**

◆ **Cafeterias**

◆ **Institutions**

◆ **Other sites outside the home.**

Due to changes in our eating habits and more knowledge about food safety hazards, recommendations for safe food handling are always changing. For example, food establishments used to cook ground meat to an internal temperature of 140°F (60°C). But that was before Shiga toxin-producing *E. coli* bacteria came on the scene. Now food establishments are required to cook ground meat to an internal temperature of 155°F (68°C) for 15 seconds. This higher temperature is needed to destroy the Shiga toxin-producing *E. coli* bacteria that may be present in the raw, ground meat. The safety of unpasteurized juices, like apple cider and fresh-squeezed orange juice, was not a concern in the past. Now we read about outbreaks in these juices involving *Salmonella, E. coli,* and parasites. As a result, most juice products are now heat pasteurized to improve their safety.

Some emerging food safety concerns include *Listeria monocytogenes* in ready-to-eat processed foods; Shiga toxin-producing *E. coli* in cut leafy greens such as spinach and lettuce; *Salmonella* in peanut butter; Hepatitis A virus in deli sandwich operations and shellfish; and parasites in fresh produce. The future may bring other types of problems.

Technologies used in the food industry are also changing. A lot of research is being done on irradiation and other types of food-processing methods that do not use heat. They may become more common in the future. You need to keep up with this information as it relates to food establishment operations.

Customers have less time to prepare food because more of them are working outside the home. As a result, they are buying more ready-to-eat food or products that require minimal preparation in the home and they are going out to eat more. These foods are produced using a variety of processing, holding, and serving methods that help protect them from contamination.

The Problem: Foodborne Illness

Foodborne illness is a disease caused by the consumption of contaminated food. A **foodborne disease outbreak** is defined as an incident in which two or more unrelated people experience a similar illness after eating a common food. Recent outbreaks of foodborne illness have been caused by:

◆ Shiga toxin-producing *E. coli* bacteria in lettuce, unpasteurized apple cider, and radish sprouts

◆ *Salmonella* spp. in cut melons, alfalfa sprouts, ice cream, roma tomatoes, raw almonds, peanut butter, and dry cereal

◆ Hepatitis A virus in raw and lightly cooked oysters and green onions

◆ *Listeria monocytogenes* in hot dogs, luncheon meats, and cheese.

A growing number of people are "at-risk" of experiencing foodborne illness.These people are frequently immunocompromized, but they do not have to be.

notes
p 302

Groups of people who are immunocompromised and at risk of foodborne illness:

◆ The very young
◆ The elderly } *easily identifiable*
◆ Pregnant or lactating women
◆ People with impaired immune systems due to cancer, AIDS, HIV, diabetes, or medications that suppress response to infection.

Foodborne illness can cause severe reactions, even death, for individuals in these at-risk groups. A safe food supply is critical to these people.

KEY TERMS

Foodborne Illness - a disease caused by the consumption of contaminated food.

Foodborne disease outbreak is an incident in which two or more unrelated people experience a similar illness after eating a common food.

Contamination

KEY TERM

Contamination - the presence of substances or conditions in food that can be harmful to humans.

Contamination is the presence of substances or conditions in food that can be harmful to humans. Raw foods can be contaminated at the farm, ranch, or on board a commercial fishing boat. Contamination can also occur as foods are handled during processing and distribution. Measures to prevent and control contamination must begin when food is harvested and continue until the food is consumed.

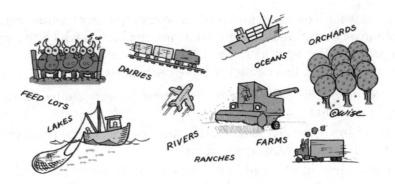

Foods can become contaminated at a variety of points as they flow from the farm to the table.

Soil, water, air, plants, animals, and humans are some of the more common sources of contamination. The contaminants that cause foodborne illness pose a special challenge because they can cause illness without changing the appearance, odor, texture, or taste of food.

Contaminants can be transferred from one food item to another by cross contamination. This typically happens when microbes from a raw food are transferred to a ready-to-eat food by contaminated hands, equipment, or utensils.

notes
p 301

Sources of Contamination

Microorganisms

Microorganisms (also called germs and microbes) are the most common types of food contamination. **Microorganisms** include:

◆ Bacteria
◆ Viruses
◆ Parasites
◆ Fungi.

These organisms are so small they can only be seen with the aid of a microscope. Bacteria and viruses are the greatest threat to food safety in food establishments. Microbes are everywhere around us – in soil, water, air, and in and on plants and animals (including humans).

Most microorganisms are harmless. However, some microbes can cause problems when they get into food. The microbes that must be controlled in a food establishment are the ones that cause foodborne illness and food spoilage. It is important to remember the microbes that cause foodborne illnesses may not alter the characteristics of the food they contaminate.

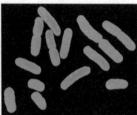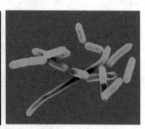
The Food Flow

Food products and the ingredients used to make them "flow" through a food establishment. The food flow begins with the purchase of safe and wholesome food and ingredients from approved sources. Once the food is delivered, it then flows through receiving into storage. The final stages in the food flow are preparation and service. These later stages include all the activities that are associated with the production, display, and service of food for your customers.

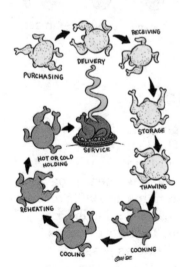

Food-handling errors during preparation and service can lead to foodborne illness. For foods that support the growth of disease-causing microorganisms, time and temperature must be monitored and controlled. Preparation of ready-to-eat foods (may be eaten without cooking or additional preparation by the customer) can also involve contact with an employee's hands or with food-contact surfaces. It is important that food employees practice good personal hygiene and wash their hands properly. Remember, during preparation and service hand washing is critical!

Preparation and service are usually the last steps before the food is eaten. Products are handled many

Microorganisms - (also called germs and microbes) are the most common types of food contamination and include bacteria, viruses, parasites, and fungi.

Contaminants that cause foodborne illness may not change the appearance, odor, texture, or taste of food.

Preparation steps frequently involve:

◆ Thawing
◆ Cooking
◆ Cooling
◆ Reheating
◆ Hot-holding
◆ Cold-holding.

times and in many different ways during this period. Learn to recognize foods that could be contaminated prior to delivery. Know how to keep these products safe until they are served or sold. The information you need is in this text or can be found in references used to prepare this program.

Facility Planning and Design

A well-planned facility with a suitable layout is essential for the smooth operation of any food establishment. Layout, design, and facilities planning directly influence:

◆ Employee safety and productivity
◆ Labor and energy costs
◆ Customer satisfaction
◆ Food safety and sanitation.

Some food establishments are located in buildings designed and constructed specifically for a food operation. Others are located in buildings converted to accommodate food preparation, handling, display, and sales. Either way, the better your facility is planned, the easier it will be to achieve your food safety goals and earn a profit.

Keeping It Clean and Sanitary

Every person working in the food industry is responsible for keeping things clean and sanitary. Effective cleaning of equipment reduces the chances of food contamination during preparation, storage, display, and service.

◆ **Cleaning** involves removal of visible soil from the surfaces of equipment and utensils
◆ **Sanitary** means healthful or hygienic.

It involves reducing the number of disease-causing microorganisms on the surface of equipment and utensils to acceptable public health levels. Something that is sanitary presents little or no risk to human health. Good sanitation minimizes attraction of pests, increases the length of time equipment is in service, improves employee morale and efficiency, and is important from other aesthetic considerations.

SOILED   CLEAN   SANITIZED

A New Approach to an Old Problem

Food industry professionals and regulatory officials agree that better ways to protect people from foodborne illness must be found. The FDA *Food Code* recommends using the Hazard Analysis Critical Control Point (HACCP) system as an important component of your food safety management program. The Pillsbury Company developed this innovative system in the 1960s for the NASA space program. Officials in the space program realized a foodborne illness in space could cause a life-threatening situation. Therefore, it was critical all food used by the astronauts be as near 100% safe as possible.

The HACCP system follows the flow of food through the food establishment and identifies each step in the process where contamination might cause the food to become unsafe. When a problem step is identified, action is taken to make the product safe or, if that is not possible, the food may have to be discarded. The HACCP system should be designed to accommodate the types of products served and the production equipment and processes used in the establishment. A more detailed discussion of the HACCP system as a component of a total food safety management program is presented in Chapter 8 of this book.

Accident Prevention and Crisis Management

Accident prevention programs are necessary in every food establishment. The cost of accidents may mean the difference between profit and loss. Beyond financial responsibilities, the loss of an employee's skills can disrupt operations and cause additional stress on other employees. Assuring a safe work environment for employees and customers requires continuous monitoring, but the rewards make it worth the effort.

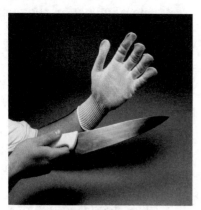

Prevention is the key to avoiding accidents.

Human error will always be a factor in food establishments, but training and proper equipment can help employees avoid accidents. Oven doors left open on deck ovens, poorly stacked boxes in storage areas, spilled food, and many other situations can lead to serious injuries. For example, a fall on a wet floor can cause expenses running into the thousands of dollars for lost time, employee compensation, and medical care.

In food establishments, another kind of crisis occurs when water supplies, electricity, or sewer systems are disrupted. The FDA *Food Code* contains specific instructions on how to operate when basic services to an establishment are lost. Public health departments can also offer assistance when you are not sure how to proceed.

Storms, tornados, hurricanes, or floods cause conditions that require changes to maintain food safety. A good disaster plan is invaluable in times of need. Managers and supervisors are expected to know how to handle these and other emergencies when they occur.

Education and Training Are Key to Food Safety

The person in charge and employees must know the correct way to manage food safety and sanitation. The importance of teaching employees about food safety is increased by the global nature of our food supply. Control of factors during growth, harvest, and shipping is not always possible when food is produced in so many different parts of the world. Also, an error in time and temperature management, cross contamination, or personal health and hygiene of food employees can increase the risk of foodborne illness. Proper storage, preparation, holding, display, and handling procedures are critical in the prevention of foodborne illness. Employees do not typically come to the job knowing this information. They have to be trained.

The prevention of foodborne illness begins with:

◆ The knowledge of where contaminants come from;

◆ How they get into food, and;

◆ What can be done to control or eliminate them.

The Role of Government in Food Safety

Government regulation in food safety oversees the food-producing system to protect the food we eat. Governmental agencies enforce laws and rules to protect food against adulteration and contamination. Regulatory personnel monitor both the process and the product to assure the safety of our food.

There are several federal regulatory agencies, such as the U.S. Food and Drug Administration (FDA) and U.S. Department of Agriculture (USDA) that set food safety regulations to make our food supply safer. The federal agencies are usually involved with assuring the safety of foods that are processed or prepared and then transported by interstate commerce. Most food establishments are governed by state or local agencies such as a health department or department of agriculture. These agencies are likely to have the most impact on the everyday activities of food establishments.

The FDA *Food Code*

In 1993, the FDA published the *Food Code* which replaced three earlier model codes. The FDA has made a commitment to revise the *Food Code* every four years, with the assistance of experts from state and local government, industry, professional associations, and colleges and universities. In this text, the *2009 Food Code* will serve as the resource from which safe food-handling recommendations are taken and will be referred to as the *Food Code* from this point forward.

The *Food Code* is not a law. Rather, it is a set of recommendations designed for use as a model by state, local, and tribal jurisdictions when formulating their own rules and regulations. You will need to obtain a copy of the health rules and regulations that apply to the food establishments in your jurisdiction.

The Role of the Food Industry in Food Safety

The food industry is assuming greater responsibility for overseeing the safety of its own processes and products. Customers expect and deserve food that is safe to eat. If a food establishment is involved in a foodborne disease outbreak, customers may take their business elsewhere or seek legal action. Financial loss and damaged reputation are two of the outcomes of a foodborne disease outbreak that can cause serious harm to the establishment found responsible for the problem. One way to prevent the harmful effects of a foodborne disease outbreak is to start a food safety management program in the food establishment. This helps assure that proper safeguards are used during food production, handling, and display. The ability to prove a food safety system was in place at the time a foodborne disease outbreak occurred is very important. It has been deemed an acceptable defense in court cases where victims of foodborne illness have sought punitive damages.

Food Protection Manager Certification

Several organizations including the Conference for Food Protection, the FDA, and the trade associations for food establishments recommend food manager certification as a way to ensure the person-in charge of a food establishments is knowledgeable about food safety.

The Conference for Food Protection has developed the standards for a Certified Food Protection Manager

A certified food protection manager is a person who:

◆ **Is responsible for identifying hazards in the day-to-day operation of a food establishment** that prepares, packages, serves, vends or otherwise provides food for human consumption

◆ **Develops or implements specific policies, procedures, or standards** aimed at preventing foodborne illness

◆ **Informs food employees and conditional food employees about their responsibility** to report to the person in charge about their health activities as they relate to diseases that are transmitted through food.

◆ **Coordinates training, supervises or directs food preparation activities and takes corrective action** as needed to protect the health of the customer

◆ **Conducts in-house, self-inspection of daily operations** on a periodic basis to see that policies and procedures concerning food safety are being followed.

A certified food protection manager must demonstrate knowledge and skills in food protection management including the following areas:

◆ Identifying foodborne illness

◆ Identifying major food allergens and the symptoms they cause

◆ Describing the relationship between time and temperature and the growth of microorganisms that cause foodborne illness

◆ Describing the relationship between employee health, personal hygiene, and food safety

◆ Describing methods for preventing food contamination at purchasing and receiving

◆ Recognizing problems and potential solutions associated with facility, equipment, and layout in a food establishment.

A more detailed list of knowledge areas is presented in Appendix E.

Some state and municipal agencies already require certification of food managers, and others are considering it. Most certification programs require managers to pass a food safety examination recognized by the Conference for Food Protection. These examinations are designed to link food safety theory to practice. By passing a certification examination, the food manager is able to demonstrate proficiency in food protection management. Information about the Conference for Food Protection and its program for recognizing food protection manager certification examinations can be found at the organization's web site at http://www.foodprotect.org

Some managers will voluntarily seek training before they take a certification examination. Some jurisdictions will require training before testing while others will require only the test. Training can involve participating in a formal course, or it could involve using a nontraditional approach such as home study, correspondence, distance learning, and interactive computer programming.

This book is written as a teaching tool to be used in conjunction with traditional and nontraditional food manager certification programs. Like the manager certification examinations, it links theory to practice. The authors of this book recognize knowing about food safety is not enough. You must also be able to apply the principles and practices that enhance food safety during preparation, display and service.

What Do You Think?

Salmonella outbreak linked to Vernon Hills Restaurant

Health officials say at least 163 people have been confirmed by laboratory analysis to be suffering from a *Salmonella* infection. Among the victims were 27 employees of the restaurant that is believed to be the source of the outbreak.

Investigators believe the outbreak was caused by an infected food employee at the restaurant who failed to wash his hands after using the bathroom. The ill employee then spread the *Salmonella* bacteria by touching food items and equipment with contaminated hands.

How could this outbreak have been prevented?

Summary

Back to the Story...

You read about a foodborne disease outbreak caused by a food employee who was infected with *Salmonella* bacteria. The bacteria were spread by fecal to oral contamination. In other words, by not using good handwashing procedures after going to the toilet, the employee did not remove the *Salmonella* bacteria from his or her hands. The bacteria were then transferred to the food and equipment by contact with the employee's soiled hands. This outbreak could have been avoided with proper hand washing and by preventing the infected food employee from working with food and food-contact surfaces.

Everyone who works in a food establishment must understand that foodborne illness is preventable.

It is the duty of every food establishment operator, manager, and employee to handle foods safely. Failure to do so can have a serious financial impact on the establishment and may cost employees their job.

The health and safety of customers can be protected by developing and implementing effective food safety and sanitation practices within a food establishment. In the following chapters, you will learn more about how food is contaminated and what actions are needed to prevent, control, and eliminate the agents that frequently cause foodborne illness and spoilage.

Quiz 1 (Multiple Choice)

Please choose the BEST answer to the questions.

1. Which of the following groups is **not** especially susceptible to foodborne illness?

    a. Young children.
    b. College students.
    c. People with weakened immune systems.
    d. Pregnant or lactating women.

2. The cost of foodborne illness can occur in the form of:

    a. Medical expenses.
    b. Loss of sales.
    c. Legal fees and fines.
    d. All of the above.

3. CDC reports show that in most foodborne illness outbreaks, mishandling of the suspect food occurred within which of the following stages?

    a. Transportation.
    b. Food establishments.
    c. Food manufacturing.
    d. Food production (farms, ranches, etc.).

4. If a utensil is sanitary it:

    a. Is free of visible soil.
    b. Has been sterilized.
    c. Is a single-service item.
    d. Has had disease-causing germs reduced to safe levels.

5. How does the FDA *Food Code* affect individual states and jurisdictions?

    a. The *Food Code* is a federal law that must be enforced by state agencies.
    b. The *Food Code* regulates food manufacturing facilities (processors) in state jurisdictions.
    c. It provides a model for new laws and rules in state and local jurisdictions.
    d. It validates current practices.

6. The event that occurs when two or more unrelated people become ill after eating the same food item is a foodborne:

    a. Illness.
    b. Contamination.
    c. Hazard.
    d. Disease outbreak.

7. Common sources of contamination for food include:

    a. Pests such as insects, rodents, and birds.
    b. Uncooked meat and poultry.
    c. Soiled hands and equipment
    d. All of the above.

Answers to the multiple-choice questions are provided in **Appendix A**.

References/Suggested Readings

Centers for Disease Control and Prevention.2000. *Surveillance for Foodborne Disease Outbreaks—United States, 1993-1997.* U.S. Department of Health and Human Services. March 17, 2000. Atlanta, GA.

Food and Drug Administration. 2009. *2009 Food Code.* U.S. Public Health Service, Washington, D.C.

Jay, James M., Martin J. Loessner, and David A. Golden. 2005. *Modern Food Microbiology*, 7th ed. Springer Publishing Co. New York, NY.

Mead, Paul S., Laurence Slutsker, Vance Dietz, Linda F. McCaig, Joseph S. Bresee, Craig Shapiro, Patricia M. Griffin, and Robert V. Tauxe. 1999. *"Food-Related Illness and Death in the United States." Emerging Infectious Diseases,* Vol. 5, No. 5, September-October, 1999. Centers for Disease Control and Prevention, Atlanta, GA.

Suggested Web Sites

Gateway to Government Food Safety Information

  www.foodsafety.gov

U.S. Department of Agriculture

  www.usda.gov

Centers for Disease Control and Prevention

  www.cdc.gov

Food and Drug Administration

  www.fda.gov

USDA/FDA Food and Nutrition Information Center

  www.nal.usda.gov/fnic

National Restaurant Association

  www.restaurant.org

Partnership for Food Safety Education

  www.fightbac.org

The Food Marketing Institute

  www.fmi.org

Conference for Food Protection

  www.foodprotect.org

Notes

2 Hazards to Food Safety

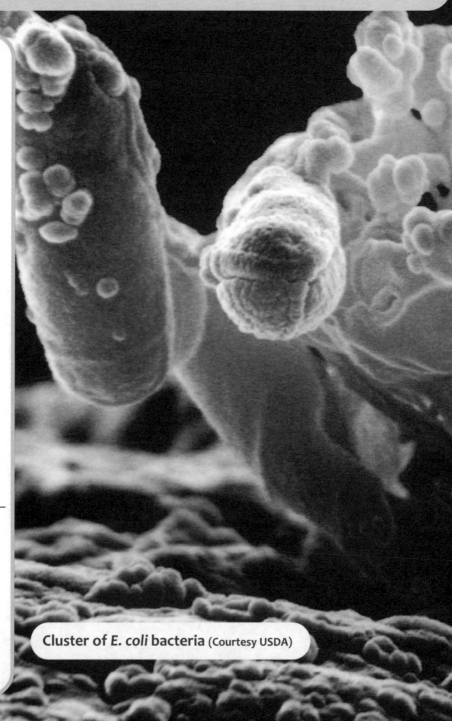

Cluster of *E. coli* bacteria (Courtesy USDA)

Key Terms

- Acidic
- Aerobic
- Alkaline
- Anaerobic
- Bacteria
- Binary fission
- Biological hazards
- Chemical hazards
- Facultative anaerobic
- FATTOM
- Food allergen
- Foodborne hazard
- Infection
- Intoxication
- Onset time
- Parasites
- Pathogenic bacteria
- pH
- Physical hazards
- Potentially hazardous foods [time/temperature control for safety foods – PHF(TCS)]
- Ready-to-eat (RTE) foods
- Spoilage bacteria
- Spore
- Toxin-mediated infection
- Vegetative cells
- Virus
- Water Activity (A_w)

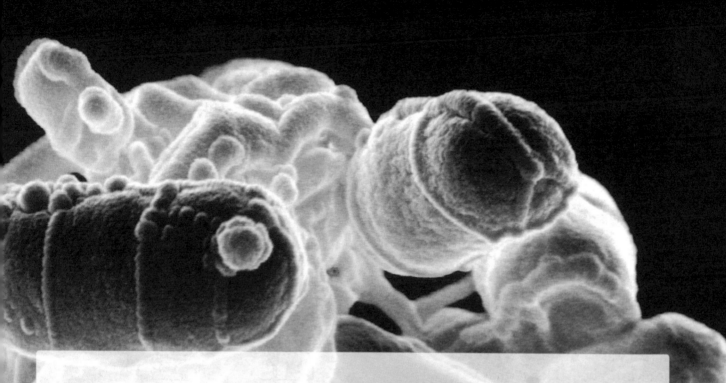

There are many different types of foodborne hazards that a food establishment may encounter which include biological, chemical, and physical hazards. Control and prevention of foodborne hazards in a food establishment start with understanding the different types of foodborne hazards.

Learning Points:

◆ List the three main categories of foodborne hazards.

◆ Identify the difference between infections, intoxications, and toxin-mediated infections as classes of foodborne illness.

◆ List the factors that promote the growth of disease-causing bacteria.

◆ Explain how temperatures in the danger zone between 41°F (5°C) and 135°F (57°C) can affect bacterial growth.

◆ List the major types of potentially hazardous foods (time-temperature control for safety foods).

◆ Identify the characteristics common to potentially hazardous foods (time-temperature control for safety foods).

What Do You Think?

Norovirus outbreak associated with restaurant employees

The local health department was notified of gastrointestinal illness from several customers after eating homemade seafood salad at the Brookville Restaurant. An investigation was initiated by the health department to identify the source and agent of infection and to determine the scope of illness among customers and employees of this national chain restaurant. Norovirus was detected in fecal specimens submitted by multiple customers and employees that worked in the restaurant. In total, over 350 restaurant customers became ill during a 3-day period. The restaurant was closed for over a week during the investigation. Upon further investigation, the local health department learned two food employees (a line cook and a server) had been sick, showing symptoms of vomiting and loose stools. The cook had vomited at home that day and then vomited again into a trash can beside the frontline workstation. An inspection of the restaurant identified deficiencies with employee handwashing practices, cleaning and sanitizing of food and non-food contact surfaces, and maintenance of hand sinks.

What went wrong in this situation and how could this outbreak have been prevented?

Foodborne Illness

Many people have had foodborne illness and not even known it. In fact, the CDC estimates that 1 in 4 Americans (76 million people) get a foodborne illness each year. The symptoms of foodborne illness are very similar to those associated with the "flu". The type of microbe, how much contamination is in the food, and the general condition of the affected person determines the severity of the symptoms of the illness.

Ingredients for foodborne illness

General symptoms of foodborne illness usually include one or more of the following:

- Headache
- Nausea
- Vomiting
- Dehydration
- Abdominal pain
- Diarrhea
- Fatigue
- Fever.

Foodborne illness is generally classified as a **foodborne infection, intoxication, or toxin-mediated infection.** Your awareness of how different microbes cause foodborne illness will help you understand how they contaminate food.

Classifications of foodborne illness

Infection: Caused by eating food that contains living, disease-causing microorganisms.

Intoxication: Caused by eating food that contains a harmful chemical or toxin produced by bacteria or other source.

Toxin-Mediated Infection: Caused by eating a food that contains harmful microorganisms that produce a toxin once inside the human intestinal tract.

KEY TERM

Onset time - the period between the time a person eats contaminated food and when they show the first symptoms of the disease.

Foodborne illnesses have different onset times. The onset time is the period between the time a person eats contaminated food and when they show the first symptoms of the disease.

Anyone can become ill from eating contaminated foods. In most cases, healthy adults will have flu-like symptoms and recover in a few days. The risks and dangers associated with foodborne illness are much more serious for people who are immunocompromised. This "at-risk" group includes infants and young children, the elderly, pregnant women, individuals with suppressed immune systems (due to cancer, diabetes, etc.), and people taking certain types of medications. For these individuals, the symptoms and duration of foodborne illness can be much more severe—even life threatening.

Immunocompromised people

At Risk Population Groups

Notes p. 302

Onset times vary depending on factors such as the victim's:

◆ Age
◆ Health status
◆ Body weight
◆ Amount of contaminant ingested with the food.

Foodborne hazard - a biological, chemical, or physical hazard that can cause illness or injury when consumed along with the food.

Biological hazards - bacteria, viruses, parasites, and fungi (yeast and mold).

Chemical hazards - toxic substances that may occur naturally or may be added during the processing of food.

Physical hazards - hard or soft foreign objects in food that can cause illness and injury.

Foodborne Hazards

A **foodborne hazard** is a biological, chemical, or physical hazard that can cause illness or injury when consumed along with the food.

Biological hazards include bacteria, viruses, parasites, and fungi (yeast and mold) and are:

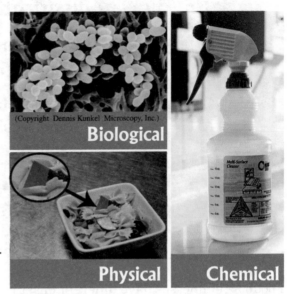

(Copyright Dennis Kunkel Microscopy, Inc.)

Biological

Physical

Chemical

◆ Very small and can only be seen with the aid of a microscope
◆ Commonly associated with animals, plants, humans, contaminated water, air, and with raw food products
◆ The most common cause of foodborne illness
◆ The primary target of a food safety program.

Chemical hazards are toxic substances that may occur naturally or may be added during the processing of food. Examples of chemical contaminants include agricultural chemicals (i.e., pesticides, fertilizers, and antibiotics), cleaning compounds, heavy metals (lead and mercury), food additives, and food allergens for allergen-sensitive people. Harmful chemicals have been associated with severe poisonings and allergic reactions. Chemicals and other nonfood items should be labeled clearly and never placed near food items.

Physical hazards are hard or soft foreign objects in food that can cause illness and injury. They include items such as fragments of glass, metal shavings, stones, toothpicks, jewelry, adhesive bandages, and human hair. These hazards result from accidental contamination and poor food-handling practices that can occur at many points in the food chain from the farm to the customer.

Bacteria - single-celled microorganisms that require food, moisture, and specific temperatures, acidity, and oxygen levels to multiply in foods.

Vegetative state - the active state of a bacterium in which the cell takes in nourishment, grows, and produces wastes.

Spore - the inactive or dormant state of some bacteria.

Bacteria

Bacteria are single-celled microorganisms that require food, acidity, temperature, time, oxygen, and moisture to multiply in foods. Bacteria can cause foodborne infections, intoxications, and toxin-mediated infections. In food establishments, most bacteria are destroyed or their growth is controlled by:

◆ Monitoring time and temperature
◆ Good personal hygiene practices
◆ An effective cleaning and sanitation program
◆ Measures that minimize cross contamination.

All bacteria exist in a "**vegetative state**." Vegetative cells grow, reproduce, and produce wastes just like other living organisms. Some bacteria have the ability to form structures called "spores." Spores help bacteria survive when their environment is too hot, cold, dry, acidic, or when there is not enough food. Spores are not able to grow or reproduce.

Keep spores from changing into the dangerous vegetative state where they can grow and cause illness.

However, when conditions become suitable for growth, a spore can "germinate" much like a seed. The bacterial spores can then convert to the vegetative state and begin to grow. Bacteria can survive for many months as spores, and it is much harder to destroy bacteria when they are in a spore form.

Spoilage and Disease-causing Bacteria

Bacteria are classified as either spoilage or pathogenic (disease-causing) microorganisms.

Spoilage bacteria break down foods so they look, taste, and smell bad. They reduce the quality of food to unacceptable levels. **Pathogenic bacteria** are disease-causing microorganisms that can make people ill and can cause death, if the vegetative bacterial cells or their toxins are consumed with food. Both spoilage and pathogenic bacteria must be controlled in food establishments.

Bacterial Growth

Bacteria reproduce when one bacterial cell divides to form two new cells. This process is called **binary fission**. The reproduction of bacteria and an increase in the number of organisms is referred to as bacterial growth. Under ideal conditions, bacteria can multiply by doubling in numbers every 15-30 minutes! That's why it is so important to control food-handling conditions and prevent bacteria from growing.

KEY TERMS

Spoilage bacteria - bacteria that break down foods so they look, taste, and smell bad.

Pathogenic bacteria - disease-causing microorganisms that can make people ill and can cause death, if the vegetative bacterial cells or their toxins are consumed with food.

Binary fission - the process where bacteria reproduce when one bacterial cell divides to form two new cells.

What Disease-causing Bacteria Need in Order to Multiply

Disease-causing bacteria need 6 conditions in order to grow and multiply:

Food - Acid - Temperature - Time - Oxygen - Moisture

An easy way to remember the 6 conditions required for the growth of disease-causing bacteria is by using the acronym F-A-T-T-O-M.

KEY TERM

FATTOM - the acronym used to indicate the six conditions bacteria need for growth– food, acid, temperature, time, oxygen, and moisture.

P. 303 notes

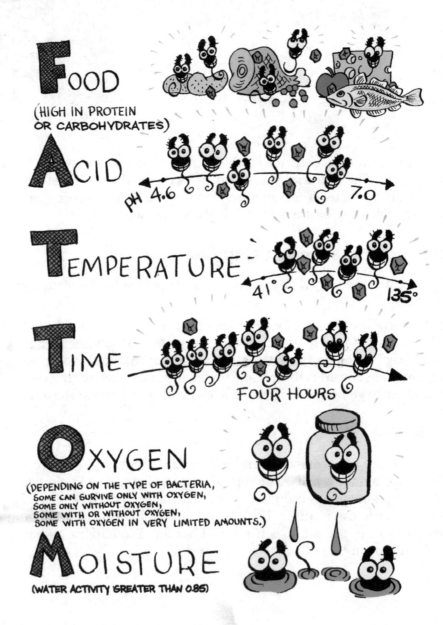

FOOD
(HIGH IN PROTEIN OR CARBOHYDRATES)

ACID
pH 4.6   7.0

TEMPERATURE
41°   135°

TIME
FOUR HOURS

OXYGEN
(DEPENDING ON THE TYPE OF BACTERIA, SOME CAN SURVIVE ONLY WITH OXYGEN, SOME ONLY WITHOUT OXYGEN, SOME WITH OR WITHOUT OXYGEN, SOME WITH OXYGEN IN VERY LIMITED AMOUNTS.)

MOISTURE
(WATER ACTIVITY GREATER THAN 0.85)

Since many foods naturally contain microorganisms, it is necessary to control one or more of these 6 conditions to prevent bacteria from multiplying.

Food

A suitable food supply is the most important condition needed for bacterial growth. Most bacteria prefer foods high in protein or carbohydrates like meats, poultry, seafood, dairy products, and cooked rice, beans, and potatoes.

Acidity

The **pH** symbol is used to designate the level of acidity or alkalinity of a food. You measure pH on a scale that ranges from 0 to 14.

Most foods are acidic and have a pH less than 7.0. Foods highly **acidic** (pH below 4.6), like lemons, limes, and tomatoes, will not normally support the growth of disease-causing bacteria. Pickling vegetables and meats preserves the food by adding acids, such as vinegar, to them. The vinegar lowers the pH of the food and stops bacterial growth.

A pH above 7.0 indicates the food is **alkaline**. Only a few foods are alkaline. Examples of alkaline foods are olives, egg whites, and soda crackers.

Most bacteria that can cause foodborne illness prefer a neutral environment (pH of 7.0) but are capable of growing in foods that have a pH in the range of 4.6 to 9.0.

Disease-causing bacteria grow best when the foods they live in and on have a pH of 4.6 or higher.

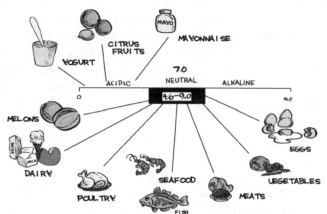

Temperature

Most disease-causing bacteria can grow within a temperature range of 41°F (5°C) to 135°F (57°C). This is commonly referred to as the food "temperature danger zone," or TDZ. A few disease-causing bacteria, such as *Listeria monocytogenes*, can grow at temperatures below 41°F (5°C), but the rate of growth is very slow.

notes
p. 303

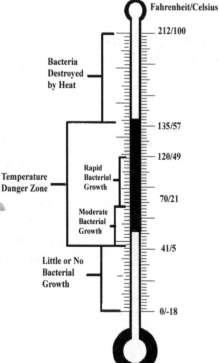

Time

For many bacteria, a single cell can generate over one million new cells in just a few hours. Because bacteria have the ability to multiply rapidly, it does not take long before many cells are produced. Bacteria need about 4 hours to grow to high enough numbers to cause illness. This includes the total time a food is between 41°F (5°C) and 135°F (57°C).

Time	0	15 min.	30 min.	60 min.	4 hrs.
# of cells	1	2	4	16	> 65,000

Bacterial cells can double in number every 15 minutes under ideal conditions.

> **Careful monitoring of time and temperature is the *most effective* way to control the growth of pathogenic and spoilage organisms.**

Oxygen

Bacteria also differ in their requirements for oxygen. **Aerobic** bacteria need about 18-21% oxygen in order to grow. This is the amount of oxygen normally present in air that we breath. **Anaerobic bacteria** cannot survive when oxygen is present because oxygen is toxic to them. These bacteria grow well in foods where oxygen has been removed such as vacuum-packaged foods or canned foods. Anaerobic conditions also exist in the internal portions of cooked food masses, such as in large stockpots, baked potatoes, or in the middle of a roast or ham. **Facultative anaerobic** forms of bacteria can grow with or without oxygen.

Moisture

Moisture is an important factor in bacterial growth. The amount of water available in food to support bacterial growth is called **water activity**. It is designated with the symbol A_w. Water activity is measured on a scale from 0.0 to 1.0. Water activity is a measure of the amount of water not bound to the food and is, therefore, available to support bacterial growth.

For example, fresh chicken has 60% water by volume, and its A_w is approximately 0.98. The same chicken, when frozen, still has 60% water by volume but its A_w is nearly 0.0. Lowering the water activity of foods to 0.85 or below prevents disease-causing organisms from growing. Drying foods, or adding salt or sugar, reduces the amount of available water. For example, jams and jellies that contain a lot of sugar have an A_w much less than 0.85. The sugar binds onto the water making it not available for bacterial growth. This intervention alone prevents the growth of disease-causing microorganisms.

KEY TERMS

Aerobic bacteria - bacteria that need about 18-21% oxygen in order to grow.

Anaerobic bacteria - bacteria that cannot survive when oxygen is present because oxygen is toxic to them.

Facultative anaerobic bacteria - bacteria that can grow with or without oxygen.

Water activity (A_w) - the amount of water in food available to support bacterial growth.

Reducing the amount of moisture in a product can prevent disease-causing organisms from growing. Moisture can be reduced by:

◆ Drying foods
◆ Adding salt
◆ Adding sugar
◆ Freezing food.

Sundried raisins and cranberries

1.0

- Dairy products
- Poultry and eggs
- Meats
- Fish and shellfish
- Cut melons and sprouts
- Steamed rice and pasta

Disease-causing bacteria can only grow in foods that have a water activity higher than 0.85.

.85

- Dry noodles
- Dry rice and pasta
- Flour
- Uncut fruits and vegetables
- Jams and jellies
- Solidly frozen foods

0

Water activity (A$_w$) of some foods sold in food establishments

Understanding the factors influencing the growth of disease-causing bacteria can help you put controls into place to prevent foodborne illness.

FOOD

ACID

TEMPERATURE

TIME

OXYGEN

MOISTURE

Potentially
hazardous food
(time/temperature
control for safety
food) [PHF (TCS)] - a food
that requires both time
and temperature control
to limit growth of disease-
causing microorganisms or
toxin formation in foods.

Potentially Hazardous Food (Time/Temperature Control for Safety Food)

Potentially hazardous food (time/temperature control for safety food), or PHF (TCS), is a term used to refer to a food that requires both time and temperature control to limit growth of disease-causing microorganisms or toxin formation in foods. Examples of common PHF (TCS) items include animal foods that are raw or heat treated, and plant foods that are heat treated or consist of raw seed sprouts, cut melons, garlic-in-oil mixtures that are not modified to inhibit the growth of disease-causing microbes, cut leafy greens, and, cut tomatoes including sliced, diced, chopped, and pureed tomatoes.

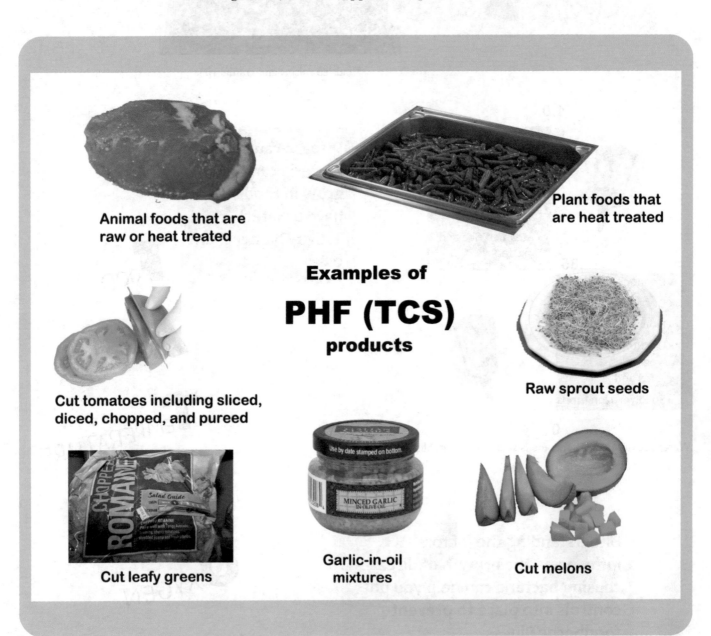

Animal foods that are raw or heat treated

Plant foods that are heat treated

Examples of

PHF (TCS)

products

Raw sprout seeds

Cut tomatoes including sliced, diced, chopped, and pureed

Cut leafy greens

Garlic-in-oil mixtures

Cut melons

PHF (TCS) takes into consideration the pH of the food, the water activity (Aw) of the food, the interaction of pH and water-activity, as well as the type of heat treatment and packaging to determine whether the food requires time and temperature control for safety. In some foods, it is possible that neither the pH value nor the water activity value is low enough by itself to control or eliminate growth of disease-causing bacteria; however, the interaction of pH and water activity together may be sufficient to inhibit microbial growth.

Determination of PHF (TCS) products has been facilitated by describing the general interactions of pH and water activity in Annex 3, Tables A and B, of the *Food Code.* These same tables are provided in Appendix B of this book.

The PHF (TCS) approach will provide more flexibility for the food industry to establish safe storage conditions for foods that will not support growth of pathogenic microorganisms or toxin formation. The concept behind PHF (TCS) helps us better understand that even a slight change in pH or water activity could make a big difference in the potential for disease-causing microorganisms to grow.

According to Food Code:

The *Food Code* has identified certain foods that are **not** PHF (TCS), and these include:

◆ Air-cooled hard-boiled eggs with shell intact, pasteurized egg with shell intact treated to destroy all viable salmonellae

◆ Commercially sterile shelf-stable foods in unopened hermetically sealed containers

◆ A food that because of its pH or water activity, or interaction of pH and water activity, is designated as a non-PHF (TCS) product

◆ A food that has undergone a product assessment showing that the growth of pathogenic microorganisms or toxin formation is unlikely due to intrinsic factors of the food (i.e., natural or added preservatives or antimicrobials), extrinsic factors of the food (i.e., reduced oxygen packaging), or a combination of these factors.

If PHF (TCS) are held in the temperature danger zone [between 41°F (5°C) and 135°F (57°C)] for 4 hours or more, infectious and toxin-producing microbes can grow to dangerous levels. PHF (TCS) have been associated with most foodborne disease outbreaks. It is critical to control the handling and storage of PHF (TCS) to prevent growth of disease-causing bacteria.

PHF (TCS) have been associated with most foodborne disease outbreaks.

Ready-to-Eat Foods

Ready-to-eat (RTE) foods can become contaminated if not handled properly. The main concern about RTE foods is that the customer may not cook or heat the product before consuming. It is important that safe food-handling practices are in place to avoid contamination of RTE foods.

Food Code
U.S. Public Health Service
FDA
2009

According to Food Code:

The *Food Code* identifies the following types of foods as ready-to-eat:

◆ Raw animal foods that are cooked (i.e., rotisserie chicken) or frozen (i.e., sushi)
◆ Raw fruits and vegetables that are washed
◆ Fruits and vegetables that are cooked for hot-holding
◆ All PHF (TCS) that are cooked and then cooled
◆ Bakery items such as bread, cakes, pies, fillings, or icing for which further cooking is not required for food safety
◆ Substances derived from plants such as spices, seasonings, and sugar
◆ Plant foods for which further washing, cooking, or other processing is not required for food safety, and from which rinds, peels, husks, or shells, if naturally present, are removed
◆ Dry, fermented sausages (i.e., dry salami or pepperoni), salt-cured meat and poultry products (i.e., prosciutto ham, country cured ham, and Parma ham), and dried meat and poultry products (i.e., jerky or beef sticks) produced in accordance with USDA guidelines and have been treated to destroy pathogens
◆ Thermally processed low-acid foods (i.e., smoked fish or meat) packaged in hermetically sealed containers.

Foodborne Illness Caused by Bacteria

Biological hazards are important for the food establishment manager to control because they lead to the majority of foodborne illness. Biological hazards, including bacteria, viruses, and parasites, are the most common agents that lead to foodborne illness. Bacteria are classified as sporeforming and non-sporeforming organisms.

Bacteria

Bacillus cereus
Campylobacter jejuni
Clostridium perfringens
Clostridium botulinum
Listeria monocytogenes
Salmonella spp.
Shiga toxin-producing
    Escherichia coli
Shigella spp.
Staphylococcus aureus
Vibrio spp.

Viruses

Hepatitis A
Norovirus
Rotavirus

Parasites

Anisakis spp.
Cryptosporidium parvum
Cyclospora cayetanensis
Giardia lamblia
Toxoplasma gondii
Trichinella spiralis

Common biological hazards in food establishments

Foodborne Illness Caused by Sporeforming Bacteria

A spore is a special structure that enables a bacterial cell to survive environmental stress such as cooking, freezing, high-salt conditions, drying, and high-acid (pH equal to or less than 4.6) conditions.

Spores are not harmful if ingested, except in a baby's digestive system where *Clostridium botulinum* spores can cause a disease called infant botulism. It is often recommended that parents avoid serving honey to babies due to the possible presence of *Clostridium botulinum* spores. If conditions in the food are suitable for bacterial growth and the spore has the ability to convert into a vegetative cell, the vegetative cell can grow in the food and cause illness if eaten.

Sporeforming bacteria are generally found in foods grown in soil, like vegetables and spices. They are most common in dried foods like dried vegetables. They may also be found in animal products. They can be particularly troublesome in food establishments when foods are not cooled properly.

For example, a 10-gallon pot of chili was prepared for the next day's salad bar display. All the ingredients (beans, meat, spices, tomato base) were mixed together and cooked to a rapid boil. Vegetative cells will die, but the more heat resistant spores may survive.

The chili was then stored in the 10-gallon pot and allowed to cool overnight in a walk-in refrigerator. It can take the core temperature of the chili 2 to 3 days to cool from 135°F (57°C) to 41°F (5°C)! If given enough time at the right temperature during the cooling process, sporeforming bacteria that survived the cooking process can change into vegetative cells and begin to grow and/or produce toxin.

FOOD WITH VEGETATIVE CELLS AND SPORE CELLS.

ONLY SPORES SURVIVE WHEN FOOD IS HEATED TO 165°F (74°C).

SPORES MAY BECOME VEGETATIVE IF FOOD IS NOT COOLED PROPERLY.

VEGETATIVE CELLS GROW AND PRODUCE TOXIN.

 SPORE   **VEGETATIVE CELL**   **TOXIN**

Spores are most likely to convert into the dangerous vegetative state when:

◆ They are "heat-shocked" during cooking which can allow the spores to convert into vegetative cells

◆ Optimum conditions exist in the food for growth (high in protein or carbohydrates, high moisture, pH greater than 4.6)

◆ Temperatures are in the food temperature danger zone or between 41°F (5°C) to 135°F (57°C) for 4 or more hours.

In the following sections of this chapter, the most common types of biological hazard are described, the common foods and route of transmission are identified, and preventive strategies are discussed.

Sporeforming bacteria

◆ Generally found in foods grown in soil, like vegetables and spices

◆ Most common in dried foods like dried vegetables

◆ May be found in animal products

◆ Foods must be cooled properly to avoid activating spores.

It is critical hot food temperatures be maintained at 135°F (57°C) or above, cold foods should be held at 41°F (5°C) or below, and foods be cooled from 135°F (57°C) to 41°F (5°C) properly.

Always cool foods as rapidly as possible to avoid activiating spores.

Bacillus cereus

Causative Agent	◆ *Bacillus cereus*
Type of Illness	◆ Bacterial intoxication or toxin-mediated infection
Symptoms	◆ **Diarrheal type:** diarrhea, abdominal cramps ◆ **Vomiting type:** vomiting, abdominal cramps
Onset	◆ **Diarrheal type:** 8 to 16 hours; usually lasts 12 to 14 hours ◆ **Vomiting type:** 30 minutes to 6 hours; usually lasts 30 minutes to 6 hours
Common Foods	◆ **Diarrhea type:** meats, milk, vegetables, fish ◆ **Vomiting type:** rice, starchy foods, grains, cereals
Commonly Caused By	◆ Improperly stored (cooled, hot-held) foods permitting spores to convert to vegetative cells. Vegetative cells grow and produce toxin in food that leads to illness.
Prevention	◆ Properly cook and hold at 135°F (57°C), cool rapidly to below 41°F (5°C), and reheat foods to 165°F (74°C).

Cooked rice

Bacillus cereus is a sporeforming bacterium that can survive with or without oxygen. It has been associated with two very different types of illnesses: one vomiting, the other diarrheal. Illness due to *Bacillus cereus* is most often attributed to foods improperly stored (cooled, hot-held), permitting the conversion of spores to vegetative cells. Vegetative cells then grow and produce toxin in the food that leads to illness.

> ***Bacillus cereus* can cause two different types of illnesses, one with vomiting and one with diarrhea.**

Clostridium perfringens

Causative Agent	◆ *Clostridium perfringens*
Type of Illness	◆ Bacterial toxin-mediated infection
Symptoms	◆ Intense abdominal pain and severe diarrhea
Onset	◆ 8 to 22 hours
Common Foods	◆ Spices, gravy, improperly cooled foods (especially meats and gravy dishes)
Commonly Caused By	◆ PHF (TCS) that have been temperature abused [not kept above 135°F (57° C) during hot-holding; or not cooled quickly enough
Prevention	◆ Properly cook, cool, and reheat foods.

Gravy in mashed potatoes

Clostridium perfringens is a nearly anaerobic (can only tolerate very little oxygen for growth), sporeforming bacteria that causes foodborne illness. PHF (TCS) that have been temperature abused [not kept above 135°F (57°C) during hot holding or not cooled quickly enough] are frequently associated with this illness. *Clostridium perfringens* causes illness due to a toxin-mediated infection where the ingested cells colonize and then produce a toxin in the human intestinal tract. Foods must be cooked to 145°F (63°C) or above. Cooked foods must be cooled from 135°F (57°C) to 70°F (21°C) within 2 hours and from 135°F (57°C) to 41°F (5°C) within a total of 6 hours. Foods must also be reheated to 165°F (74°C) within 2 hours and held at 135°F (57°C) until served.

Prevent *Clostridium perfringens* by cooling foods properly.

Clostridium botulinum

Causative Agent	◆ *Clostridium botulinum*
Type of Illness	◆ Bacterial intoxication
Symptoms	◆ Dizziness, double vision, difficulty in breathing and swallowing, headache
Onset	◆ 12 to 36 hours; usually lasts several days to a year
Common Foods	◆ Low-acid foods (pH above 4.6), inadequately heat-processed and packaged anaerobically (metal can or vacuum pouch), and held in the food temperature danger zone. Some types of this organism can grow in temperatures as low as 38°F (3°C). Examples: home-canned green beans, meats, vacuum packaged fish, and garlic or onions stored in oil and butter respectively.
Commonly Caused By	◆ Eating foods that were not heat-processed correctly and packaged anaerobically.
Prevention	◆ Properly heat-process and cool vacuum-packaged and other reduced-oxygen packaged foods ◆ DO NOT use home-canned foods.

Salmon

Clostridium botulinum is an anaerobic (cannot tolerate oxygen), sporeforming bacterium that causes foodborne intoxication due to improperly heat-processed foods, especially home canning. Do not can food products in a food establishment. The organism produces a neurotoxin as it grows that is one of the deadliest biological toxins known to man. This toxin is not heat stable and can be destroyed if the food is boiled for about 20 minutes. However, botulism still occurs because people do not want to boil food that has already been cooked. Illness due to *Clostridium botulinum* is almost always attributed to ingestion of foods that were not heat processed correctly and packaged anaerobically.

Foodborne Illness Caused by Non-sporeforming Bacteria

Bacteria that cannot form spores are called non-sporeforming bacteria. These types of bacteria can only exist as a vegetative cell. Compared to bacterial spores, vegetative cells are easily destroyed by proper cooking. There are numerous examples of non-sporeforming foodborne bacteria that are important in the food industry, including *Salmonella* spp., *Listeria monocytogenes,* and *E. coli.*

Campylobacter jejuni

Causative Agent	◆ *Campylobacter jejuni*
Type of Illness	◆ Bacterial infection
Symptoms	◆ Watery, bloody diarrhea
Onset	◆ 2 to 5 days; usually lasts 2 to 7 days
Common Foods	◆ Raw poultry, raw milk, raw meat
Commonly Caused By	◆ Cross contamination of surfaces between raw poultry and other foods ◆ Failure to wash hands after handling raw poultry and meat
Prevention	◆ Properly handle and cook raw meats and poultry, properly clean and sanitize food-contact surfaces and properly wash hands.

Most reports indicate that *Campylobacter jejuni* is either the No. 1 or No. 2 cause of bacterial foodborne infection in the United States. This organism needs between 3 to 6% oxygen to grow. The organism grows well in the hair follicles of chickens and in fluid milk. Some reports state that *Campylobacter jejuni* may be found in nearly 100% of raw chicken. The organism is often transferred from raw meats to other foods by cross contamination, typically from a food-contact surface (such as a cutting board or knife) or a food employee's hands.

DID YOU KNOW?

Campylobacter jejuni is one of the top 2 causes of bacterial foodborne infection in the United States.

Shiga toxin-producing *Escherichia coli*

Causative Agent	◆ Shiga toxin-producing *Escherichia coli*
Type of Illness	◆ Bacterial infection or toxin-mediated infection; at special risk are children up to 16 years old and the elderly
Symptoms	◆ Bloody diarrhea followed by kidney failure and hemolytic uremic syndrome (HUS) in severe cases
Onset	◆ 12 to 72 hours; usually lasts from 1 to 3 days
Common Foods	◆ Raw and undercooked beef and other red meats (especially ground meats), raw finfish, improperly pasteurized milk, unpasteurized apple cider, lettuce and other leafy greens
Commonly Caused By	◆ Employees failing to wash hands after using restroom ◆ Cross contamination between soiled equipment and surfaces
Prevention	◆ Practice good food sanitation, hand washing ◆ Properly handle and cook ground meats to an internal temperature of at least 155°F (68°C) for 15 seconds; prevent cross contamination and keep hot foods above 135°F (57°C) and cold foods below 41°F (5°C). Use only pasteurized apple cider or fruit juice and milk products.

The *Escherichia coli* (or *E. coli*) group of bacteria includes five different types of foodborne pathogens. The most important type is enterohemorrhagic *E. coli* called Shiga toxin-producing *E. coli*. These facultative anaerobic (grow with or without oxygen) bacteria can be found in the intestines of warm-blooded animals, especially cows. The illness caused by Shiga toxin-producing *E. coli* can be an infection or a toxin-mediated infection. Only a small amount of bacteria is required to produce an illness (as little as 10 cells). As a result, the organism may not need to grow in foods; it may just need to survive. Non-PHF (TCS) have been linked to illnesses including apple cider (pH = 4.0).

Shiga toxin-producing *E. coli* is usually transferred to foods such as beef through contact with the intestines of slaughtered animals. For example, apple cider was likely contaminated on the farm from apples obtained from orchards where cattle grazed. Transmission can occur within a food establishment if employees who are carriers do not wash their soiled hands properly after going to the toilet. Cross contamination by soiled equipment and utensils may also spread Shiga toxin-producing *E. coli*.

DID YOU KNOW?

It only takes a small amount of *E. coli* bacteria to become ill, so the bacteria just need to survive. Even non-PHF (TCS) products have been linked to *E. coli* outbreaks.

Listeria monocytogenes

Causative Agent	◆ *Listeria monocytogenes*
Type of Illness	◆ Bacterial infection
Symptoms	◆ Healthy adult: flu-like symptoms ◆ Highly susceptible populations: septicemia, meningitis, encephalitis, birth defects ◆ Unborn fetus: stillbirth
Onset	◆ 1 day to 3 weeks; indefinite duration depending on when treatment is administered
Common Foods	◆ Raw meats, raw poultry, dairy products, cooked luncheon meats and hot dogs, raw vegetables, protein-based salads, seafood
Commonly Caused By	◆ Cross contamination ◆ Improperly cooked foods
Prevention	◆ Properly store and cook foods, avoid cross contamination, rotate processed refrigerated foods using FIFO to ensure timely use.

Listeria monocytogenes is a facultative anaerobic (grow with or without oxygen) bacterium that causes foodborne infection. This microbe is important to food establishment operations because it has the ability to survive under many environmentally stressful conditions such as in high-salt foods and, unlike most other foodborne pathogens, can grow at refrigerated temperatures below 41°F (5°C). Transmission to foods can occur by cross contamination by people or equipment or if foods are not cooked properly.

Listeria monocytogenes can grow at refrigerated temperatures.

Salmonella spp.

Causative Agent	◆ *Salmonella* spp.
Type of Illness	◆ Bacterial infection
Symptoms	◆ Nausea, fever, vomiting, abdominal cramps, diarrhea
Onset	◆ 6 to 48 hours; usually lasts 2 to 3 days
Common Foods	◆ Raw meats, raw poultry, eggs, fresh fruits and vegetables, milk, dairy products, pork
Commonly Caused By	◆ Failure of employees to wash hands ◆ Cross contamination of raw foods, especially poultry
Prevention	◆ Properly cook foods; example: *Salmonella* bacteria will be destroyed when poultry is cooked to an internal temperature of 165°F (74°C) for 15 seconds and when eggs are cooked to 145°F (63°C) for 15 seconds. Clean and sanitize raw food-contact surfaces after use; make sure food employees wash their hands adequately before working with food, avoid cross contamination.

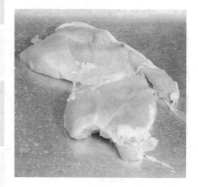

Salmonella are facultative anaerobic (grow with or without oxygen) bacteria frequently implicated as a foodborne infection. *Salmonella* are found in the intestines of warm-blooded animals and humans. It frequently gets into foods as a result of fecal contamination or cross contamination. Transmission to foods is commonly through cross contamination where fecal material is transferred to food through contact with raw foods (especially poultry), contaminated food-contact surfaces (i.e., cutting boards), or infected food employees.

Sprouts and melons are a common source for *Salmonella* spp.

Shigella spp.

Causative Agent	◆ *Shigella* spp.
Type of Illness	◆ Bacterial toxin-mediated infection
Symptoms	◆ Bacillary dysentery, diarrhea, fever, abdominal cramps, dehydration
Onset	◆ 1 to 7 days; duration depends on when treatment is administered
Common Foods	◆ Foods prepared with human contact: ready-to-eat salads (i.e., potato, chicken), raw vegetables, milk, dairy products, raw poultry, nonpotable water, ready-to-eat meat
Commonly Caused By	◆ Infected employee working with food ◆ Employee failing to wash hands after using restroom and before working with food
Prevention	◆ Wash hands and practice good personal hygiene, properly cook foods, avoid cross contamination, wash produce and other foods with potable water (water that is safe to drink). Do not allow individuals who have been diagnosed with shigellosis to handle food.

Shigella spp. are facultative anaerobic bacteria (grow with or without oxygen) that account for about 10% of foodborne illnesses in the United States. These organisms come mainly from the feces of humans and cause the infection, shigellosis. The bacterium produces a toxin during the infection that causes watery diarrhea. Water contaminated by human fecal material and food/utensils handled by employees who are carriers of the bacteria can cause this problem. Illness from *Shigella* spp. is most often attributed to contaminated ready-to-eat foods handled by an infected food handler.

Shigella spp. is most often attributed to foods prepared with human contact.

DID YOU KNOW?

Shigella spp. accounts for about 10% of all foodborne illnesses in the United States.

No crops

Staphylococcus aureus

Causative Agent	◆ Staphylococcus aureus
Type of Illness	◆ Bacterial intoxication
Symptoms	◆ Nausea, vomiting, abdominal cramps, headaches
Onset	◆ 1 to 6 hours, usually 2 to 4 hours; usually lasts 1 to 2 days
Common Foods	◆ Foods prepared with human contact: cooked ready-to-eat foods such as luncheon meats, ready-to-eat meat, deli salads (such as taco, potato, egg, and tuna salads), meat, poultry, custards, high-salt foods such as ham, and milk and dairy products, processed foods
Commonly Caused By	◆ Contaminated employee hands ◆ Employees transferring germs to food by improperly using tasting spoons and ladles ◆ Time and temperature abuse
Prevention	◆ Wash hands and practice good personal hygiene, avoid coughing and sneezing near food, do not reuse tasting spoons and ladles, properly clean and bandage cuts, burns or wounds on hands and wear plastic gloves. Cooking WILL NOT inactivate the toxin.

Staphylococcus aureus is a facultative anaerobic bacterium (grow with or without oxygen) that produces a heat-stable toxin as it grows on or in foods. This bacterium can also grow on cooked, and otherwise safe, foods recontaminated by food employees who mishandle the food. These bacteria are commonly found on human skin, hands, hair, and in the nose and throat. They may also be found in burns, infected cuts and wounds, pimples, and boils. These organisms can be transferred to foods easily, and they can grow in foods that have a high salt or high sugar content, and have a lower water activity. They grow well in a high-salt concentration environment, such as on hams and luncheon meats. *Staphylococcus aureus* bacteria do not compete well when other types of microorganisms are present. However, they grow well when alone and without competition from other microbes, such as in high-salt foods. Foods requiring considerable food preparation and handling are especially susceptible. The bacteria are also spread by droplets of saliva from talking, coughing, and sneezing near food. Food employees who improperly use tasting spoons and ladles can transfer bacteria from their mouth to food. Contaminated human hands combined with temperature abuse usually cause most problems associated with *Staphylococcus aureus*.

DID YOU KNOW?
Staphylococcus aureus bacteria are naturally found on the skin and hair and in the nose and throat of humans, even healthy people.

Vibrio spp.

Causative Agent	◆ Vibrio spp.
Type of Illness	◆ Bacterial intoxication
Symptoms	◆ Headache, fever, chills, diarrhea, vomiting, severe electrolyte loss, gastroenteritis
Onset	◆ 2 to 48 hours
Common Foods	◆ Raw or improperly cooked fish and shellfish
Commonly Caused By	◆ Undercooked seafood ◆ Raw seafood
Prevention	◆ Practice good sanitation, properly cool foods, implement procedures to separate raw and ready-to-eat seafood display cases, buy seafood from approved sources only.

There are three organisms within the *Vibrio* group of bacteria connected with foodborne infections. They include *Vibrio cholera*, *Vibrio parahaemolyticus*, and *Vibrio vulnificus*. All are important since they are very resistant to salt and are common in seafood.

Since the organism is inherent in many types of raw seafood, transmission to other foods by cross contamination is a concern. Most illnesses are caused by the consumption of raw or undercooked seafood.

Shell Stock tags

Undercooked seafood is the most common source of *Vibrio* spp.

Foodborne Illness Caused by Viruses

Viruses are now thought to be the No. 1 cause of foodborne and waterborne diseases in the United States. The viruses that cause foodborne disease differ from foodborne bacteria in several ways. Viruses are much smaller than bacteria, and they require a living host (human or animal) to replicate. Viruses do not multiply in foods. However, a susceptible person needs to consume only a few viral particles in order to experience an infection.

Viruses are usually transferred from one food to another, from a food employee to a food, or from a contaminated water supply to a food. A PHF (TCS) is not needed to support survival of viruses. The viruses of primary importance to food establishments are Noroviruses and Hepatitis A virus. Proper hand washing and separation of raw and ready-to-eat foods are important keys to controlling the spread of foodborne viruses.

Viruses

Proper hand washing and separation of raw and ready-to-eat foods can help prevent the spread of foodborne viruses.

Norovirus

Causative Agent	◆ Norovirus
Type of Illness	◆ Viral infection
Symptoms	◆ Vomiting, diarrhea, abdominal pain, headache, low-grade fever
Onset	◆ 24 to 48 hours, usually lasts 1 to 3 days
Common Foods	◆ Sewage-contaminated water; contaminated salad ingredients; raw clams, oysters; foods contaminated by infected food employees
Commonly Caused By	◆ food handling by infected employees ◆ contaminated water ◆ contaminated salad ingredients
Prevention	◆ Use potable water; cook all shellfish; handle food properly; meet time, temperature guidelines for PHF (TCS); practice good personal hygiene and wash hands and fingernails thoroughly; keep raw and ready-to-eat seafood products separate.

Norovirus is the most common agent of foodborne illness in the United States, causing an estimated 23 million cases each year. The virus is primarily transmitted by ingestion of food and water contaminated with feces that contain the Norwalk virus.

DID YOU KNOW?

Norovirus is the most common cause of foodborne illness in the United States, with over 23 million cases each year.

Hepatitis A virus

Causative Agent	◆ Hepatitis A virus
Type of Illness	◆ Viral infection
Symptoms	◆ Fever, nausea, vomiting, abdominal pain, fatigue, swelling of the liver, jaundice
Onset	◆ 15 to 50 days; a mild case usually lasts several weeks, more severe cases can last several months
Common Foods	◆ Raw and lightly cooked oysters and clams harvested from polluted waters; raw vegetables that have been irrigated or washed with polluted water; foods prepared with contact by infected employee, including salads, sliced luncheon meats, salad bar items, sandwiches, bakery products; contaminated water
Commonly Caused By	◆ Food handled by infected employees ◆ Eating raw seafood from polluted waters
Prevention	◆ Buy clams, oysters, and molluscan shellfish from approved sources; keep raw and ready-to-eat foods separate during storage and display; handle foods properly and cook them to recommended temperatures; wash hands and practice good personal hygiene.

Hepatitis A - A long-lived virus

An employee can harbor the Hepatitis A virus for up to 6 weeks and not show symptoms of illness.

Food employees are contagious for 1 week before the symptoms appear, and for 2 weeks after the symptoms do appear.

Hepatitis A virus is a foodborne virus associated with many foodborne infections. Hepatitis A virus causes a liver disease called infectious hepatitis. The Hepatitis A virus is a particularly important hazard to food establishments because employees can harbor the virus for up to 6 weeks and not show symptoms of illness. Food employees are contagious for 1 week before onset of symptoms and 2 weeks after the symptoms of the disease appear. During that time, infected employees can contaminate foods and other employees by spreading fecal material from unwashed hands and nails. Hepatitis A virus is very hardy and can live for several hours in a suitable environment. The virus is transmitted by ingestion of food and water that contain the Hepatitis A virus. Raw seafood and foods handled by infected human hands are the largest threat of transmission and disease from Hepatitis A virus.

Proper hand washing is critical to prevent the transmission of Hepatitis A.

Foodborne Illness Caused by Parasites

Foodborne parasites are another important foodborne biological hazard. **Parasites** are small or microscopic creatures that need to live on or inside a living host (usually animal or human) to survive. Many parasites can enter the food system and cause foodborne illness. Parasites cause food infections. Parasitic infection is far less common than bacterial or viral foodborne illnesses. In this chapter, we list a few of the most troublesome ones that may appear in food establishments.

KEY TERM

Parasites - small or microscopic creatures that need to live on or inside a living host (usually animal or human) to survive.

Anisakis spp.

Causative Agent	◆ *Anisakis* spp.
Type of Illness	◆ Parasitic infection
Symptoms	◆ Coughing if worms attach in throat, vomiting and abdominal pain if worms attach in stomach, sharp pain and fever if worms attach in large intestine
Onset	◆ 1 hour to 2 weeks
Common Foods	◆ Raw or undercooked seafood; especially bottom-feeding fish
Commonly Caused By	◆ Eating undercooked or raw seafood
Prevention	◆ Cook fish to the proper temperature throughout, freeze to meet *Food Code* specifications, inspect seafood and handle carefully, purchase seafood from approved supplier.

Anisakis spp. are nematodes (roundworms) associated with foodborne infection from consumption of contaminated fish. The worms are about 1 to 1-1/2 inches long and the diameter of a human hair. They are beige, ivory, white, gray, brown, or pink. Other names for this parasite are "cod worm" (not to be confused with common roundworms found in cod) and "herring worm." The natural hosts of the parasite are walruses, and perhaps, sea lions and otters. The worms are transferred to fish, their intermediate host, in the water in which the walruses live. Humans become the accidental host upon eating fish infested with the parasites. Humans do not make good hosts for the parasites. The worms will not complete their life cycles in humans, and eventually die.

Bottom-feeding fish such as salmon are a common source of *Anisakis* spp.

Cyclospora cayetanensis

Causative Agent	◆ *Cyclospora cayetanensis*
Type of Illness	◆ Parasitic infection
Symptoms	◆ Watery and explosive diarrhea, loss of appetite, bloating
Onset	◆ Usually within 1 week; symptoms persist for weeks or months if untreated
Common Foods	◆ Contaminated water, strawberries, raspberries, and fresh produce
Commonly Caused By	◆ Foods contaminated on the farm ◆ Improper handwashing by an infected employee
Prevention	◆ Good sanitation and personal hygiene, purchase foods from reputable supplier.

Black raspberries

Cyclospora cayetanensis is a parasite that has been reported more frequently in the past few decades. *Cyclospora* usually finds its way into water and then can be transferred to foods. Foods usually become contaminated after coming in contact with fecal material from polluted water or an infected food employee. The most common outbreaks of cyclosporiasis have been associated with fresh fruits and vegetables that were contaminated at the farm. *Cyclospora* is passed from person to person by fecal-oral transmission The *Cyclospora* parasite may take days or weeks after a person eats a contaminated food to become infectious.

The most common outbreaks of *Cyclospora* have been associated with fresh fruits and vegetables contaminated at the farm.

Trichinella spiralis

Causative Agent	◆ *Trichinella spiralis*
Type of Illness	◆ Parasitic infection from a nematode worm
Symptoms	◆ Early symptoms: nausea, vomiting, diarrhea, sweating, abdominal pain; in later stages: fever, swelling of tissues around eyes, muscle stiffness
Onset	◆ 2 to 28 days; death may occur in severe cases
Common Foods	◆ Primarily wild game meats (bear, walrus) and occasionally undercooked pork products
Commonly Caused By	◆ Undercooked game meats and pork
Prevention	◆ Cook foods to the proper temperature throughout, i.e., no pink color in cooked pork products.

Trichinella spiralis is a foodborne roundworm that causes a parasitic infection. It must be eaten with the infected fleshy muscle of certain meat-eating animals to be transmitted to a new host. Meat-eating, scavenger animals carry this parasite. These animals are exposed to the parasite when they eat infected tissues from other animals and garbage that contains contaminated raw-meat scraps.

Cook game meat and raw pork to proper temperatures.

Cryptosporidium parvum

Causative Agent	◆ *Cryptosporidium parvum*
Type of Illness	◆ Parasitic infection
Symptoms	◆ Severe watery diarrhea
Onset	◆ Within 1 week of ingestion
Common Foods	◆ Contaminated water, food contaminated by infected food employees
Commonly Caused By	◆ Contaminated water supply ◆ Failure of employees to wash hands after going to the toilet
Prevention	◆ Use potable water supply; practice good personal hygiene and hand washing.

Cryptosporidium parvum is a parasite found in water that has been contaminated with cow feces. The parasite causes foodborne infection and is considered an important source of nonbacterial diarrhea in the United States. The parasite could occur, theoretically, on any food touched by a contaminated food handler. It is primarily transmitted by a water supply contaminated with feces and by fecal contamination of food and food-contact surfaces. Parasite prevention starts with providing a potable water supply in the food establishment and handling foods carefully to prevent contamination and cross contamination. Food employees must practice good personal hygiene and wash hands thoroughly before working with food and after going to the toilet.

Fresh strawberries can be a source of *Cryptosporidium parvum*.

Problems Caused by Fungi

Yeast and molds make up the group of microbes called fungi. Yeasts and molds mainly contribute to food spoilage. When yeast grows in a food, it produces gases and alcohol. This may cause the food to taste or smell or the package to swell. Yeasts do not lead to foodborne illness. In fact, yeasts are often used to produce fermented foods such as beer, wine, and cheeses.

Molds usually grow in foods low in moisture (such as bread or cheese), acidic foods (fruit juices), or foods that are high in sugar (jams and jellies). They often appear very colorful or "cotton-like" or "powdery" in appearance. Like yeasts, molds themselves do not cause foodborne illness. However, if they grow long enough on foods, they can produce a substance called a "mycotoxin."

Mycotoxins can cause foodborne illness and some cancers. Mycotoxins are usually considered a type of chemical hazard, and, these toxins may not be destroyed by cooking. Several foods including peanuts, grains, and corn are tested for mycotoxins at the farm and by manufacturers.

Foodborne Illness Caused by Chemicals

Chemical hazards are usually classified as either naturally occurring or man-made chemicals. Naturally occurring chemicals include toxins produced by a biological organism. Man-made chemicals include substances added, intentionally or accidentally, to a food during processing. A summary of some of the more common naturally occurring and man-made chemicals is provided below.

Types of chemical hazards in a food establishment

Naturally Occurring:

◆ Allergens

◆ Ciguatoxin

◆ Mycotoxins

◆ Scombrotoxin

◆ Shellfish toxins.

Man-made Chemicals:

◆ Cleaning solutions

◆ Food additives

◆ Pesticides

◆ Heavy metals.

(Source: *Food Code*)

Naturally Occurring Chemicals

Food Allergens

According to the *Food Code,* over 12 million Americans suffer from one or more food allergies. Between 5 and 8% of children and 1 to 2% of adults are allergic to certain chemicals in foods and food ingredients. These chemicals are commonly referred to as food allergens. A **food allergen** is a protein that causes a person's immune system to overreact. Some common symptoms of food allergies are hives; swelling of the lips, tongue, and mouth; difficulty breathing or wheezing; and vomiting, diarrhea, and cramps. These symptoms can occur in as little as 5 minutes. In severe situations, a life-threatening allergic reaction called "anaphylaxis" can occur. Anaphylaxis is a condition that occurs when many parts of the body become involved in the allergic reaction. Symptoms of anaphylaxis include itching and hives; swelling of the throat and difficulty breathing; and lowered blood pressure and unconsciousness.

KEY TERM

Food allergen - a protein that causes a person's immune system to overreact.

The FDA has identified the major source of food allergens as the "Big Eight" from the 170 different foods known to cause allergic reactions. These 8 allergens are considered to be "major food allergens." The only way for a person who is allergic to one of these foods to keep from having an allergic reaction is to avoid eating the food containing the allergen. In many cases, it doesn't take much of the food to produce a severe reaction. As little as half a peanut can cause a severe reaction in highly sensitive people. Like other chemical toxins, cooking will not inactivate the allergen.

The "Big Eight" major food allergens

- Milk
- Eggs
- Wheat
- Peanuts
- Soybeans
- Tree nuts (i.e., almonds and pecans)
- Fish
- Crustacean shellfish (i.e., lobster and shrimp)

Anaphylaxis

A person who has a severe allergy may have a life-threatening reaction called anaphylaxis when they ingest even a small portion of the allergen.

Symptoms include:
- Itching and hives
- Swelling of the throat and difficulty breathing
- Lowered blood pressure
- Unconsciousness.

If you notice these symptoms, get this person medical treatment immediately!

Allergies can be very serious. You need to be able to identify which foods contain these ingredients. The FDA now requires certain ingredients to be listed on the label of packaged foods. Always read label information to determine if food allergens may be present.

The Food Allergy and Anaphylaxis Network (FAAN) urges food establishments to take customer food allergy requests and questions seriously. FAAN advises food establishments to always let customers make their own informed decision about what foods to buy and eat. FAAN recommends when a customer informs a food employee that someone has a food allergy, they follow the following four Rs:

- *Refer* the food allergy concern to the manager or person in charge.
- *Review* the food allergy with the customer and check ingredient labels.
- *Remember* to check the preparation procedure for potential cross-contact.
- *Respond* to the customer and inform him or her of your findings.

If a customer has an allergic reaction, call 911 and assist emergency medical personnel dispatched to attend to the allergic person.

The responsibility of the foodservice industry is to assure appropriate label warnings are given to potential purchasers. Retailers have special responsibilities in regard to foods sold from open containers (i.e., on delicatessen counters). When such foods contain a major food allergen, the warning should be clearly displayed by the foods in question. For both the retail and foodservice industry, staff must be trained to take great care to avoid cross contamination such as might occur when using the same utensil or other handling equipment for a food containing a food allergen and one that does not contain it.

Manufacturers are required to list the ingredients in their products.

When one of the "Big Eight" allergens is present, it should be listed as well.

Ciguatoxin

Causative Agent	◆ Ciguatoxin
Type of Illness	◆ Fish toxin originating from toxic algae of tropical waters
Symptoms	◆ Vertigo, nausea, hot/cold flashes, diarrhea, vomiting, shortness of breath
Onset	◆ 30 minutes to 6 hours; usually lasts a few days but death can occur from concentrated dose of toxin
Common Foods	◆ Marine finfish including grouper, barracuda, snapper, jack, mackerel, triggerfish, reef fish
Commonly Caused By	◆ Eating contaminated fish
Prevention	◆ Purchase fish from a reputable supplier; cooking WILL NOT inactivate the toxin.

Ciguatoxin poisoning is an example of an intoxication caused by eating contaminated tropical reef fish. The toxin is found in tiny, free-swimming sea creatures called "algae" that live among certain coral reefs. When small reef fish eat the toxic algae, it is stored and it accumulates in the flesh, skin, and organs. When bigger fish such as barracuda eat the small reef fish such as mackerel, mahi-mahi, bonito, jackfish, and snapper, the toxin accumulates even more in the flesh and skin of the consuming fish. The toxin does not affect the contaminated fish. The toxin is heat stable and it is not destroyed by cooking. There is no commercially known method to determine if ciguatoxin is present in a particular fish.

Ciguatoxin can be found in marine finfish, such as red snapper.

Scombrotoxin

Causative Agent	◆ Scombrotoxin
Type of Illness	◆ Seafood toxin originating from histamine- producing bacteria
Symptoms	◆ Dizziness, burning feeling in the mouth, facial rash or hives, shortness of breath, peppery taste in mouth, headache, itching, teary eyes, runny nose
Onset	◆ Few minutes to 30 minutes; recovery usually occurs in 8 to 12 hours
Common Foods	◆ Tuna, mahi-mahi, bluefish, sardines, mackerel, anchovies, amberjack, abalone, Swiss cheese
Commonly Caused By	◆ Eating foods that contain histamine
Prevention	◆ Purchase fish from a reputable supplier; store fish between 32°F (0°C) and 39°F (4°C) to prevent growth of histamine-producing bacteria; toxin IS NOT inactivated by cooking.

Eating foods high in a chemical compound called histamine causes scombrotoxin, also called histamine poisoning. Histamine is usually produced by certain bacteria when they decompose foods containing a protein called "histidine." Dark meat of fish has more histidine than other fish meat. Histamine is not inactivated by cooking. Over time, bacteria inherent to a particular food can break down histidine and cause the production of "histamine," which causes reactions when people consume the contaminated product. Leaving fish out at room temperature is a common way for histamine to be produced in affected foods.

Scombrotoxin can be found in fish such as raw tuna.

Shellfish Toxins – PSP, DSP, DAP, NSP

Causative Agent	◆ Shellfish toxins produced by certain algae called dinoflagellates
Type of Illness	◆ Intoxication
Symptoms	◆ Numbness of lips, tongue, arms, legs, neck; lack of muscle coordination
Onset	◆ 10 to 60 minutes
Common Foods	◆ Contaminated mussels, clams, oysters, scallops
Commonly Caused By	◆ Eating contaminated shellfish harvested from polluted waters.
Prevention	◆ Purchase shellfish only from a reputable supplier, avoid buying shellfish harvested by sport fishermen or poached from polluted waters.

There are numerous other shellfish toxins that are known to cause foodborne illness including Paralytic Shellfish Poisoning (PSP), Diarrhetic Shellfish Poisoning (DSP), Domoic Acid Poisoning (DAP), and Neurotoxic Shellfish Poisoning (NSP). Many of these toxins are produced by toxic algae that are consumed by shellfish that get their nutrients by capturing their food by "filter feeding." When filter-feeding shellfish, such as mussels, clams, oysters, and scallops, feed on the toxic algae, they accumulate the toxins in their internal organs and become toxic to humans when the shellfish are consumed. The amount of toxin in the shellfish depends on the amount of toxic algae in the water and the amount of water filtered by the shellfish. Most cases of seafood toxin are caused by contaminated shellfish that have been harvested by sport fishermen or poached from polluted waters. Commercially harvested shellfish are rarely involved in foodborne, disease outbreaks because regulatory agencies monitor the level of toxin in the water and shellfish during high-risk periods (May through October).

Added Man-Made Chemicals

There is an extensive list of other chemicals that are intentionally added to foods that may pose a potential health risk. These added chemicals may include food additives, food preservatives, and pesticides. Pesticides leave residues on fruits and vegetables and can usually be removed by a vigorous washing procedure. Nonintentionally added chemicals may include contamination by chemicals such as cleaning and sanitary supplies. Also, chemicals from containers or food-contact surfaces of inferior metal that are misused may lead to heavy metal or inferior metal poisoning (cadmium, copper, lead, galvanized metals, etc.).

Employee medications can be another potential source of chemical contamination. According to the *Food Code*, only those medications necessary

Store employee medications away from food.

for an employee's health are allowed in a food establishment. This does not apply to medicines stored or displayed for sale in retail stores. Employee medications must be clearly labeled and stored in an area away from food, equipment, utensils, linens, and single-use items like straws, eating utensils, and napkins. Medications stored in a refrigerator where food is stored must be kept in a clearly labeled and covered container located on the lowest shelf of the unit. Medications cannot be stored in display cases and walk-in cooler units.

Foodborne Illness Caused by Physical Hazards

Physical hazards are foreign objects in food that can cause illness and injury. They include items such as fragments of glass, metal shavings from dull can openers, toothpicks from club sandwiches, human hair, jewelry, and bandages that may accidentally be lost by a food employee and enter food. Stones, rocks, or wood particles may contaminate raw fruits and vegetables, rice, beans, and other grain products.

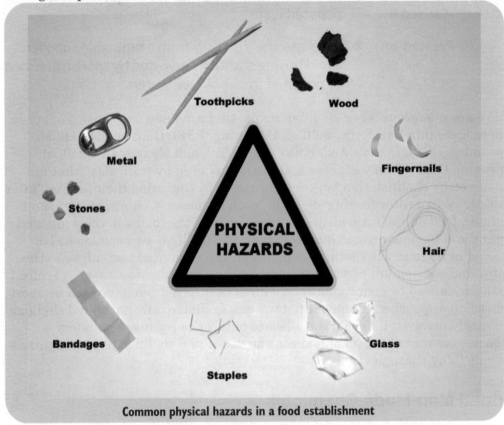

Common physical hazards in a food establishment

Physical hazards commonly result from accidental contamination and poor food-handling practices that can occur at various points in the food chain from harvest to consumer.

To prevent physical hazards:

◆ Wash raw fruits and vegetables thoroughly
◆ Visually inspect foods that cannot be washed (such as ground beef).

Food employees must be taught to handle food safely to prevent contamination by unwanted foreign objects such as glass fragments and metal shavings. Finally, food employees should not wear jewelry when involved in the production of food, except for a plain wedding band.

International Food Safety Challenges

"Mad cow disease" or bovine spongiform encephalopathy (BSE) is a fatal brain disorder that occurs in cattle and is caused by some unknown agent. There is a connection between animal feed made from the parts of sheep that carry the organisms called scrapie. Use of animal feed made from such ingredients has been banned in the United States.

The connection between BSE and humans was uncovered in Great Britain in the 1990s when several young people died of a brain disorder, a new variation of a rare problem, Creutzfeldt-Jakob disease (CJD) that typically strikes the elderly.

Efforts of the U.S. Food and Drug Administration (FDA), U.S. Department of Agriculture (USDA), the Centers for Disease Control and Prevention (CDC), and other federal organizations, including state regulatory and health agencies, have limited the diseases from occurring in this country. Advisories are regularly issued to those traveling to Great Britain and Europe about the safety of eating beef and beef products.

Foot- (hoof-) and-mouth disease is another identified problem in the world. It is not a food safety concern, but is a concern for animal health and economics. The disease is a highly contagious viral infection of cattle, sheep, goats, deer, and other cloven-hoofed animals. It causes blisters on the mouth, teats, and soft tissues of the animal's feet and mouth. The animals rarely recover. This disease is not a significant health concern for humans and was last found in the United States in 1929. In 2001, an outbreak of foot-and-mouth disease occurred in the United Kingdom (England, Scotland, Wales, Northern Ireland) and Europe. A large number of animals were destroyed to contain the infection. Affected areas were placed under quarantine. The danger from this disease is to animals, not humans; therefore, foot-and-mouth disease is not classified as a foodborne illness.

Summary

Back to the Story... The vignette at the beginning of this chapter identified norovirus as the cause of foodborne illness within a restaurant. Norovirus has been identified as the most common vehicle of foodborne illness in the United States. The primary reason this foodborne outbreak occurred is due to sick employees handling food and transferring the organism to the food. If food employees are infected with norovirus, they are not permitted to work in a food establishment until cleared by a physician, nurse practitioner, or physician assistant. This outbreak could have been prevented if proper policies had been in place at the restaurant. (cont.)

What Do You Think?

Norovirus outbreak associated with restaurant employees

The local health department was notified of gastrointestinal illness from several customers after eating homemade seafood salad at the Brookville Restaurant. An investigation was initiated by the health department to identify the source and agent of infection and to determine the scope of illness among customers and employees of this national chain restaurant. Norovirus was detected in fecal specimens submitted by multiple customers and employees that worked in the restaurant. In total, over 350 restaurant patrons became ill during a 3-day period. The restaurant was closed for over a week during the investigation. Upon further investigation, the local health department learned two food employees (a line cook and a server) had been sick, showing symptoms of vomiting and loose stools. The cook had vomited at home that day and then vomited again into a trash can beside the frontline workstation. An inspection of the restaurant identified deficiencies with employee hand-washing practices, cleaning and sanitizing of food and non-food contact surfaces, and maintenance of hand sinks.

What went wrong in this situation and how could this outbreak have been prevented?

There are many different types of foodborne hazards that a food establishment may encounter. They are classified as biological, chemical, or physical hazards. These hazards differ depending on the type of food and method of preparation involved. Food establishments are toward the end of the food production chain and this is the area where foods are prepared for consumption or sold for consumer preparation. Therefore, it is important to control and prevent foodborne hazards as much as possible to reduce the risk of foodborne illness associated within your establishment. Control and prevention of foodborne hazards in a food establishment start with understanding the different types of foodborne hazards. The next step is to understand how to control foodborne hazards with time/temperature control, good personal hygiene, cleaning and sanitation, and prevention of cross contamination. Prevention, using these four approaches, will be the focus of Chapter 3.

Quiz 2 (Multiple Choice)

Choose the BEST answer for each question.

1. Most foodborne illness is caused by:

    a. Poor quality food.

    b. Biological hazards.

    c. Unsanitary utensils.

    d. Physical hazards.

2. Which of the following groups of hazards are most likely to cause a foodborne disease outbreak?

    a. Bacteria and viruses.

    b. Parasites and molds.

    c. *Vibrio* spp. and *Shigella* spp.

    d. Chemical and physical hazards.

3. Bacteria are one of the most common causes of foodborne disease in a food establishment because:

    a. Under ideal conditions, they can grow very rapidly.

    b. Bacteria are found naturally in many foods.

    c. Bacteria can be easily transferred from one source to another.

    d. All of the above.

4. Some bacteria have the ability to survive heat, lack of moisture, cold, or acidic conditions by forming a special resistant structure called a:

   a. Vegetative cell.
   b. Parasite.
   c. Spore.
   d. Colony.

5. Bacteria grow best in the temperature danger zone which includes all temperatures between:

   a. 0°F (-18°C) and 220°F (104°C).
   b. 0°F (-18°C) and 135°F (57°C).
   c. 41°F (5°C) and 135°F (57°C).
   d. 41°F (5°C) and 220°F (104°C).

6. Bacteria that cause foodborne illness will only grow on foods that have a pH at ____ or above and a water activity (A_w) above ____.

   a. 3.2; .85
   b. 4.6; .85
   c. 6.5; .80
   d. 8.0; .70

7. Some bacteria form spores to help them:

   a. Reproduce.
   b. Move more easily from one location to another.
   c. Survive adverse environmental conditions.
   d. Grow in high acid foods.

8. Organisms that negatively affect the taste and appearance of foods, but *detect c̄ senses* do not cause foodborne illness, are called:

   a. Spoilage bacteria.
   b. Viruses.
   c. Pathogenic bacteria.
   d. Toxins.

9. Foods that contain protein or carbohydrates, a water activity of 0.85 or above, and a pH of 4.6 are called _____ foods.

   a. Shelf stable.
   b. Potentially hazardous food (time/temperature control for safety food)
   c. Nonhazardous.
   d. Spoiled.

10. Which of the following is a histamine poisoning?

   a. Ciguatoxin.
   b. Scombrotoxin.
   c. Mycotoxin.
   d. Shellfish toxin.

Answers to the multiple-choice questions are provided in **Appendix A**.

References/Suggested Readings

Adams, M. R. and M. O. Moss. 2000. *Food Microbiology*. Royal Society of Chemistry, London, England.

Heymann, David (Ed.). 2004. *Control Of Communicable Diseases in Man*, (*18th ed.*), American Public Health Association, Washington, D.C.

Doyle, Michael P., Beuchat, Larry R., and Thomas J. Montville. 1997. *Food Microbiology: Fundamentals and Frontiers*. American Society for Microbiology, Washington, D.C.

Food and Drug Administration. 2009. *FDA 2009 Food Code*. U.S. Public Health Service, Washington, D.C.

Food Protection Report. June 2002. Vol. 18, No. 6., Pike & Fishers, Silver Springs, MD.

Hackney, C., M.D. Pierson and G. Banwart. 1996. *Basic Food Microbiology*, (3rd ed). Van Nostrand Reinhold, New *York*.

Institute of Food Technologists. 2001. Evaluation and Definition of Potentially Hazardous Foods. www.cfsan.fda.gov/~comm/ift4-toc.html.

Jay, J.M., M. Loessner, and D.A. Golden. 2005. *Modern Food Microbiology*, 7th Edition. Springer Science, New York, NY.

Longrèe, K., and G. Armbruster. 1996. *Quantity Food Sanitation*. Macmillan, New York, NY.

McSwane, D., Rue, N., Linton, R. 2005. *Essentials of Food Safety and Sanitation, 4th ed.* Prentice Hall, Upper Saddle River, NJ.

Suggested Web Sites

Gateway to Government Food Safety Information
     www.foodsafety.gov

Centers for Disease Control and Prevention (CDC)
     www.cdc.gov

The Bad Bug Book
     vm.cfscan.fda.gov/~mow/intro.html

USDA/FDA Food and Nutrition Information Center
     www.nal.usda.gov/fnic/

The Food Allergy and Anaphylaxis Network
     www.foodallergy.org

The Food Marketing Institute
     www.fmi.org

National Restaurant Association
     www.nra.org

U.S. Department of Agriculture (USDA)
     www.usda.gov

Food and Drug Administration (FDA)

www.fda.gov

Environmental Protection Agency (EPA)

www.epa.gov

Partnership for Food Safety Education

www.fightbac.org

IFT Report, *Evaluation and Definition of Potentially Hazardous Foods*

http://www.cfsan.fda.gov/~comm/ift4-toc.html.

FMI "Quick Sheet" and allergens posters

www.fmi.org/foodsafety/FinalQuikSheet.pdf.

FDA Guide to Minimize Microbial Food Safety Hazards of Fresh Cut Fruits and Vegetables

http://www.cfsan.gov/~dms/prodgui4.html

Factors That Affect Foodborne Illness

Key Terms

- ◆ Calibrate
- ◆ Cross contamination
- ◆ Dial-faced bimetal thermometer
- ◆ Digital thermometer
- ◆ Food temperature–measuring device
- ◆ Personal hygiene
- ◆ Temperature and time abuse
- ◆ Thermocouple
- ◆ T-Stick type melt devices

The most effective ways to prevent foodborne illness are to keep foods at the proper temperature, use good personal hygiene, and control contamination and cross contamination. These are the essentials of safe food management.

Learning Points:
◆ Identify potential problems related to temperature and time abuse of foods.
◆ Describe how to properly measure and maintain food temperatures to assure that foods are handled safely.
◆ Identify potential problems related to a food employee's poor personal hygiene.
◆ Explain how to improve personal hygiene habits to reduce the risk of foodborne illness.
◆ Identify potential problems related to cross contamination of food.
◆ Discuss procedures and methods to prevent cross contamination.

What Do You Think?

A bad batch of potato salad

Three hundered and eighteen people were reported ill from a New Jersey beach resort community after eating potato salad purchased at a local restaurant during the 4th of July holiday. Most of the people that were sick complained of intense vomiting, diarrhea, fever, nausea, abdominal cramps and severe muscle pains. The majority of people started to become ill just a few hours after eating, and most were sick from 1-2 days. Elderly victims had symptoms that lasted up to 4-5 days.

The local health department was contacted shortly after many cases of suspected food poisoning were identified at a nearby hospital emergency room. An investigation by the health department identified the likely causative agent of foodborne disease as the toxin produced by the *Staphylococcus aureus* bacteria.

Interviews with the employees who worked at the restaurant indicated they had not had formal training in good personal hygiene practices and temperature control. The potato salad was made with cooked whole potatoes that were sliced by hand and mixed with other ingredients in a 7-gallon container and then stored in the cold room for the next day. Health department officials estimated that the cooling time, after mixing the ingredients, was over 20 hours.

How could this outbreak have been prevented? See the summary at the end of this chapter.

Agents that Lead to Foodborne Illness in the United States

In Chapter 2, we identified that the most common causes of foodborne disease are from biological hazards, like viruses and bacteria, but that illnesses can also occur from chemical and physical hazards. Each year, the Centers for Disease Control and Prevention (CDC) provide a summary of statistics related to foodborne outbreaks and food-related illnesses in the United States. The chart on the next page shows a 2006 accumulated summary for the 5-year period from 2001-2005.

Note the majority of outbreaks and illnesses are caused by Norovirus, which is often transferred to foods by human contact via the fecal to oral route. Hand washing becomes the most important preventive measure for control of this virus. *Salmonella* spp., *Clostridium perfringens*, and *Staphylococcus aureus* were the most common bacterial agents of foodborne disease. As you can see, chemical hazards are far less likely to occur with Scombroid toxin (histamine) being the most common chemical agent reported.

Factors That Contribute to Foodborne Illness

CDC is an agency of the federal government that has an important role in food safety. One of CDC's responsibilities is to collect statistics about diseases that affect people in the United States, including foodborne illness. CDC statistics show most outbreaks of foodborne disease occur because food is mishandled. As we saw in Chapter 1, the major contributors to foodborne illness are presented below.

The information below helps us focus our prevention efforts at controlling the contributing factors that are most likely to lead to foodborne illness for the hazards that are most likely to be present in the food.

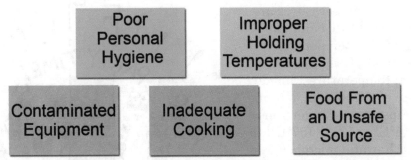

Major contributors to foodborne illness (Source: CDC-MMWR 2000)

Food-related illnesses (2001-2005)		
Hazard	**No. of Illnesses**	**% of Illnesses**
Biological		
Norovirus	9,877	38
Salmonella spp.	3,393	13
Clostridium perfringens	2,077	8
Staphylococcus aureus toxin	659	3
Shiga toxin-producing *Escherichia coli*	470	2
Campylobacter spp.	299	1
Bacillus cereus	160	1
Shigella spp.	659	3
Chemical		
Scombroid toxin (histamine)	117	<1
Ciguatoxin	59	<1
Mushroom toxins	6	<1
(source: CDC MMWR, 2006)		

What Is Temperature and Time Abuse?

KEY TERM

Temperature and time abuse - describes situations when foods are exposed to temperatures in the TDZ for enough time to allow growth of harmful microorganisms, or not cooked or reheated sufficiently to destroy harmful microorganisms.

Controlling temperature and time is perhaps the most critical way to assure food safety. Most cases of foodborne illness can in some way be linked to temperature and time abuse. The term **temperature and time abuse** is used to describe situations when foods are:

◆ Exposed to temperatures in the temperature danger zone (TDZ) for enough time to allow growth of harmful microorganisms

◆ Not cooked or reheated sufficiently to destroy harmful microorganisms.

In Chapter 2, *Hazards to Food Safety,* you learned harmful microbes can grow in PHF (TCS) when temperatures are between 41°F (5°C) and 135° F (57°C), or the "temperature danger zone." Keep the internal temperatures, inside the core of a food item, out of the temperature danger zone [41°F (5°C) to 135°F (57°C)] to prevent harmful microbes from growing. Higher temperatures used in cooking destroy most microbes, however, bacterial spores, toxins produced by microbes, and chemical toxins may not be destroyed by normal cooking temperatures.

Keep cold food temperatures below 41°F (5°C) and out of the temperature danger zone to prevent most disease-causing microbes from growing. Bacteria that can grow at lower temperatures do so very slowly.

The Temperature Danger Zone (TDZ)

135°F (57°C)

41°F (5°C)

There are unavoidable situations during food production when foods must pass through the temperature danger zone such as:

◆ Cooking
◆ Cooling
◆ Reheating
◆ Food handling and preparation (slicing, mixing and assembling).

During these activities, you must minimize the amount of time foods are in the temperature danger zone to control the extent of microbial growth. When it is necessary for a food to pass through the temperature danger zone, do it as quickly as possible. In addition, foods should pass through the danger zone as few times as possible.

Cooking and reheating foods are two very important processes for safe food management.

In addition to destroying microorganisms, heating foods improves food texture and flavor. As you learned in Chapter 2, many raw foods naturally contain harmful microbes or can easily become contaminated during handling. When you cook and reheat foods properly, microbes are reduced to safe levels or are destroyed. Cooking and reheating foods are two very important processes for safe food management.

Measuring Food Temperatures Correctly

Maintaining safe food temperatures is an essential and effective part of food safety management. You must know *how* to measure food temperatures correctly to prevent temperature abuse. Thermometers, thermocouples, and other temperature-measuring devices are used to measure temperature of stored, cooked, hot-held, cold-held, and reheated foods. The following chart shows different types of temperature-measuring devices and their features.

A temperature-measuring device with a small diameter probe must be available to measure the temperature of thin foods such as meat patties and fish filets.

Dial-faced, metal stem type (bimetallic)

- Most common type of temperature-measuring device used in food-service and retail food establishments
- Used to measure internal food temperature at every stage in the flow of food
- Measures temperatures ranging from 0°F (-18°C) to 220°F (104°C) with 2°F increments
- Stem of bimetallic thermometer must be inserted at least 2 inches into the food item being measured.

Courtesy of Cooper Instrument Corp.

Digital

- Displays the temperature numerically
- Measures a wider range of temperatures than a dial faced temperature-measuring device.

Courtesy of Cooper Instrument Corp.

Thermocouple

- Provides a digital readout of the temperature
- Has a wide variety of interchangeable probes
- Sensing portion is often at the tip of the probe.

Courtesy of Cooper Instrument Corp.

Infrared

- Measures the surface temperature of food without actually touching the food (reduces the chance of cross contamination)
- May require up to 20 minutes to adjust after use for hot and cold temperatures ("thermal shock") before use
- Accuracy must be checked frequently.

Courtesy of Raytek Corp.

T-Sticks (melt devices)

- Measure only one temperature
- Change color when indicated temperature is reached
- Used to monitor food temperatures and sanitizing temperature in dishwashing machines.

(cont.)

Courtesy of T-Stick

Courtesy of Cooper Instrument Corp.

Refrigerator/Freezer Thermometer

Built-in

◆ Used to monitor air temperature in refrigerated and frozen cases.

Portable

◆ Used to simulate the temperature of foods during cold storage

◆ Helpful for identifying cold and warm spots in reach-in and walk-in cold storage areas caused by air circulation, faulty equipment, or frequent opening and closing of the door.

Maximum Registering (holding)

◆ Used to measure the temperature of hot water used to sanitize items in mechanical dishwashing machines.

Temperature-measuring Device Guidelines:

◆ Temperature-measuring devices typically measure food temperatures in degrees Fahrenheit (denoted as °F), degrees Celsius (denoted as °C), or both.

◆ Food temperature-measuring devices scaled only in Celsius or dually scaled in Fahrenheit and Celsius must be accurate to +1.8°F (+1°C). Food temperature-measuring devices scaled in Fahrenheit only must be accurate to +2°F.

◆ Mercury-filled and glass type thermometers should not be used in food establishments.

◆ Clean and sanitize temperature-measuring devices properly to avoid contaminating food that is being tested. This is very important when testing raw and then ready-to-eat food items. To clean and sanitize a food temperature-measuring device, wipe off any food particles, place the stem or probe in sanitizing solution for at least 5 seconds, then air-dry.

◆ When monitoring only raw foods, or only cooked foods being held at 135°F (57°C), wipe the stem of the temperature-measuring device with an alcohol swab between measurements.

When and How to Calibrate a Thermometer

Before you use a thermometer as a temperature-measuring device, you need to **calibrate** it and make sure it is measuring temperature accurately. Dial-faced metal stem type (bimetal) thermometers should be calibrated:

◆ Before their first use
◆ At regular intervals
◆ If dropped or otherwise damaged
◆ If used to measure extreme temperatures
◆ Whenever accuracy is in question.

Calibrate dial-faced thermometers by the boiling point or ice point method. See specific directions for

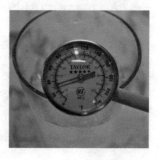

calibration as described in the following section. Use pliers or an open-ended wrench to adjust the indicator needle on the dial face.

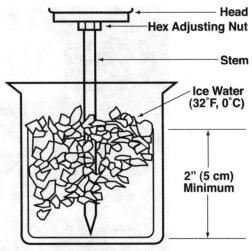

Ice Point Method

1. Insert the probe into a cup of crushed ice.
2. Add enough cold water to remove any air pockets that might remain.
3. Wait until the temperature stabilizes and adjust the needle to 32°F (0°C).

Boiling Point Method

Immerse at least the first 2 inches of the stem from the tip (the sensing part of the probe) into boiling water and adjust the needle to 212°F (100°C). At higher altitudes, the temperature of the boiling point will vary. Consult your local health department if you have any questions about the boiling point temperature in your area.

Calibration of a dial-faced thermometer using the ice point method

Source: Food Safety Inspection Service, U. S. Department of Agriculture, (www.fsis.usda.gov/Fact_sheets/kitchen_thermometers/index.asp) Nov.1, 2009

Measuring Food Temperature

The sensing portion of a food thermometer is toward the end of the stem or probe. On the bimetal thermometer, the sensing portion usually extends from the tip of the probe to the "dimple" mark that is approximately 1 inch from the tip. An average of the temperature is measured over this distance. The sensing portion for digital and thermocouple thermometers is closer to the tip of the probe.

How to Accurately and Safely Measure Food Temperatures

1. Use an approved temperature-measuring device that measures temperatures from 0°F(-18°C) to 220°F(104°C).
2. Locate the sensing portion of the measuring device.
3. If the temperature measurement device can be calibrated, calibrate it using the ice or boiling point method.
4. Clean and sanitize the probe of the temperature-measuring device according to a recommended procedure.
5. Measure the internal temperature of the food by inserting the probe into the center or thickest part of the item.
6. Always wait for the temperature reading to stabilize.

The approximate temperature of packaged foods can be measured accurately without the need to open the package. Place the stem or probe of the thermometer between two packages of food or fold the package around the stem or probe to make good contact with the packaging to get a good temperature measurement.

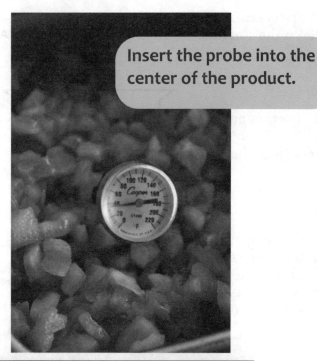

Insert the probe into the center of the product.

Wait for the temperature to stabilize before removing probe.

Measure the surface temperature of salads with an infrared thermometer.

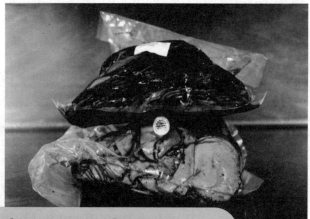

Place thermometer between packages of prepared foods to measure temperature.

Preventing Temperature and Time Abuse

Temperature control for PHF (TCS) is important in almost all stages of food preparation and handling. Measuring temperatures of PHF (TCS) is an important responsibility for all food handlers. The following chart lists temperature and time guidelines for handling food throughout the flow of food, from receiving raw materials, through food preparation, to display and service. Each of these guidelines will be emphasized throughout the remainder of this book.

Time and Temperature

Receiving and Storing:

Frozen and refrigerated receiving/storage practices prevent or slow the growth of harmful microorganisms.

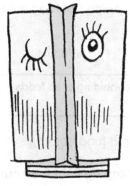

Food Product	Internal Temperature
Frozen foods	Solidly frozen
Refrigerated foods	41°F (5°C) or lower
Raw shell eggs	45° F (7° C) ambient temperature or below

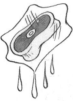

Thawing:

Take food from frozen to nonfrozen to minimize the product's time in the temperature danger zone. Keep PHF (TCS) below 41°F (5°C) at all times.

Method	Internal Temperature	Times
In refrigerator	41°F (5°C) or lower	Typically takes 2-3 days
Submerged under cool running water 70°F (21°C)	Ready-to-eat product not to exceed 41°F (5°C); water temperature not to exceed 70°F(21°C)	Thawed portions of raw animal foods requiring cooking should not be allowed to rise above 41°F (5°C) for more than 4 hours, including the time the food is being thawed and prepared for cooking

Cooking:

Safely heating a food product from raw to ready-to-eat with minimum holding times before serving.

Food Product	Minimum Internal Temperature	Times
*Meat roast (rare)	130°F (54°C)	112 minutes
	140°F (60°C)	12 minutes
Meat and pork (other than roast), fish	145°F (63°C)	15 seconds
Ground meat, mechanically tenderized meat, ground pork, ground game animals	155°F (68°C)	15 seconds
Meat roast (medium), pork roast, ham	145°F (63°C)	4 minutes
Poultry, ground poultry, stuffed meats, and stuffed food products	165°F (74°C)	15 seconds

* See additional cooking times and temperatures in FDA *Food Code* - paragraph 3-401.11(B)(2)

2009 FDA *Food Code*

Time and Temperature (continued)

Hot-holding:

Keeping hot food out of the temperature danger zone

Food Product	Internal Temperature
Hot-holding of all foods	135°F (57°C) or above

Cold food holding:

Keeping cold food out of the temperature danger zone

Food Product	Internal Temperature
Cold-holding of all foods	41°F (5°C) or below

Cooling hot foods:

Rapid reduction of temperature through and out of the temperature danger zone

Part	Internal Temperature	Times
Hot Food Cooling part 1	From 135° to 70°F (57° to 21°C)	2 hours or less
Hot Food Cooling part 2	From 135° to 41°F (57° to 5°C) or below	Within 6 hours or less

Frozen food holding:

Keeping food solidly frozen

Food Product	Internal Temperature
Frozen Food	Solidly Frozen 0°F (-18°C) recommended

Reheating:

Bringing food back up to serving temperature

Method	Internal Temperature	Times
Reheating	165°F (74°C) or above	Within 2 hours
Reheat commercially processed, intact packaged, ready-to-eat food	135°F (57°C) or above	Within 2 hours

Remember, there's NEVER been a case of foodborne illness that couldn't have been prevented!

2009 FDA *Food Code*

Keep Cold Foods Cold and Hot Foods Hot!

Frozen foods should be kept solidly frozen until they are ready to be used. Freezing helps to retain product quality and proper frozen food temperatures do not permit microorganisms to grow. Cold temperatures also help to preserve the color and flavor characteristics. Frozen foods can be stored for long periods of time without losing their wholesomeness and quality.

Refrigerated foods are held cold, not frozen. PHF (TCS) should be maintained cold at 41°F (5°C) or below. Do not forget that some harmful bacteria and many spoilage bacteria can grow at temperatures below 41°F (5°C), although their growth is very slow. By keeping cold foods at 41°F (5°C) or below, you can reduce the growth of most harmful microorganisms and extend the shelf-life of the food. For maximum quality and freshness, hold cold foods for the shortest amount of time possible.

> **Keep cold foods at 41°F (5°C) or below.**

Applying heat is another critical method used to ensure food safety. Heat food to proper temperatures to destroy harmful bacteria. Established safe cooking temperatures are based on the type of food and the method used to heat the product. Depending on the food type, foods are heated to 130°F (54°C) to 165°F (74°C) during cooking. Cooked foods, as well as those foods that have been cooled and then reheated, must be hot-held at 135°F (57°C) or above until used. You must keep foods hot to stop growth of harmful bacteria.

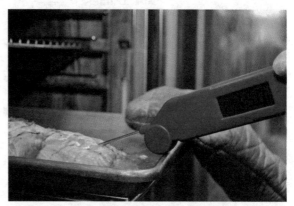

Depending on the food type, foods are heated to 130°F (54°C) to 165°F (74°C) during cooking.

There are times during food production when foods must be in the temperature danger zone. Recognize the time spent in the temperature danger zone should be minimal for PHF (TCS).

Improper holding temperatures (hot-holding, cooling, and cold- holding) is a common contributing factor that leads to foodborne illness. Spores of certain bacteria like *Clostridium botulinum, Clostridium perfringens*, and *Bacillus cereus* can survive cooking temperatures. Remember, if spores survive and are exposed to ideal conditions, they can again become vegetative cells and begin to grow in foods. Temperature and time control can prevent this.

The *Food Code* contains cooling requirements for PHF (TCS) which requires foods to be cooled from 135°F (57°C) to 70°F (21°C) within the first 2 hours of cooling, and from 135°F (57°C) to 41°F (5°C) or less within a total cooling time of 6 hours.

To destroy bacteria that may have grown during the cooling process, reheat foods to 165°F (74°C) within 2 hours to prevent the number of organisms from reaching levels that can cause foodborne illness.

> *According to Food Code:*
>
> **Food Code**
> U.S. Public Health Service
> FDA
> 2009
>
> The *Food Code* contains cooling requirements for PHF (TCS) which requires foods to be cooled from 135°F (57°C) to 70°F (21°C) within the first 2 hours of cooling, and from 135°F (57°C) to 41°F (5°C) or less within a total cooling time of 6 hours.

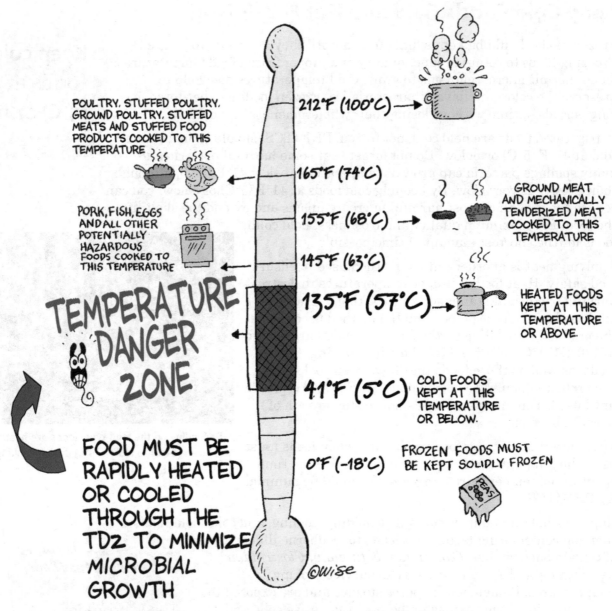

POULTRY, STUFFED POULTRY, GROUND POULTRY, STUFFED MEATS AND STUFFED FOOD PRODUCTS COOKED TO THIS TEMPERATURE

PORK, FISH, EGGS AND ALL OTHER POTENTIALLY HAZARDOUS FOODS COOKED TO THIS TEMPERATURE

TEMPERATURE DANGER ZONE

FOOD MUST BE RAPIDLY HEATED OR COOLED THROUGH THE TDZ TO MINIMIZE MICROBIAL GROWTH

212°F (100°C)

165°F (74°C)

155°F (68°C)

145°F (63°C)

135°F (57°C)

41°F (5°C)

0°F (-18°C)

GROUND MEAT AND MECHANICALLY TENDERIZED MEAT COOKED TO THIS TEMPERATURE

HEATED FOODS KEPT AT THIS TEMPERATURE OR ABOVE.

COLD FOODS KEPT AT THIS TEMPERATURE OR BELOW.

FROZEN FOODS MUST BE KEPT SOLIDLY FROZEN

@wise

Keep It Hot, Keep It Cold, or Don't Keep It!

Proper Thawing of Food

The preferred method for thawing foods is in the refrigerator at 41°F (5°C) or below. This prevents the food from entering the food temperature danger zone.

Other acceptable methods for thawing include using a microwave oven, thawing as a part of the cooking process, or submerging packaged foods under cool running water [70°F (21°C)] for a specified amount of time. Proper thawing reduces the chances for bacterial growth, especially for bacteria that may be present on the outer surfaces of food.

More detailed strategies for minimizing the amount of time a food is in the temperature danger zone during cooling, thawing, and food preparation will be presented in chapter 4 of this book.

The Importance of Good Personal Hygiene and Hand Washing

The cleanliness and personal hygiene of food employees are extremely important because people can carry disease. If a food employee is not clean, the food can become contaminated as the food handler comes in contact with it. **Good personal hygiene is essential for those who handle foods.** Examples of good personal hygiene behaviors include:

◆ Maintaining good health and reporting when a person is sick to avoid spreading possible infections.

◆ Knowing when and how to properly wash hands

◆ Wearing clean clothing

◆ Maintaining good personal habits (bathing, washing and use of hair restraints, keeping fingernails short and clean, washing hands after using toilet, etc.)

DID YOU KNOW?

Staphylococcus aureus, Hepatitis A virus, and *Shigella* spp. are examples of pathogens that may be found in and on the human body and can be transferred to foods by hand contact.

Hand Washing

Just think of all the things a food handler's hands touch during a typical workday. They may take out the trash, cover a sneeze, scratch an itch, or mop up a spill. When you touch your face or skin, run your fingers through your hair or a beard, use the toilet, or blow your nose, you transfer potentially harmful germs to your hands. Personnel involved in food preparation and service must know how and when to wash their hands. Soap, warm water, and friction are needed to adequately clean skin. A significant number of germs are removed by friction alone. The steps on the following page show the proper way to wash your hands.

Always Wash Hands:

◆ Before food preparation

◆ After touching bare human body parts, except clean hands and clean exposed arms

◆ After using the toilet

◆ After coughing, sneezing, using a handkerchief or disposable tissue, using tobacco, eating, or drinking

◆ During food preparation when switching between working with raw foods and ready-to-eat products

◆ After engaging in any activities that may contaminate hands (taking out the garbage, wiping counters or tables, handling cleaning chemicals, picking up dropped items, etc.)

◆ After caring for or touching service animals or aquatic animals.

Hand Washing

Food Code
U.S. Public Health Service
2009

◆ Hands must be washed in a separate sink specified as a handwashing sink.

◆ An automatic hand-washing facility may be used by food employees to clean their hands if the system can remove the types of soils encountered in the food operation.

◆ Each handwashing sink must be provided with hand cleanser (soap or detergent) in a dispenser and a suitable hand-drying device.

◆ If a nail brush is used, it must be kept clean and sanitary at all times.

SAFE HANDS...

1. WET HANDS

2. APPLY SOAP

3. BRISKLY RUB HANDS FOR 10-15 SECONDS

4. SCRUB FINGERTIPS AND BETWEEN FINGERS

5. SCRUB FOREARM TO JUST BELOW ELBOW

6. RINSE FOREARMS AND HANDS

7. DRY HANDS AND FOREARMS

8. TURN OFF WATER USING PAPER TOWEL

9. TURN DOORKNOB AND OPEN DOOR USING PAPER TOWEL

10. DISCARD TOWEL

...MEAN SAFE FOOD

✔ DO	⊘ DON'T
✔ The entire process of washing hands, wrists, and forearms in a handwashing sink should last at least 20 seconds.	⊘ Wash your hands in a food prep, warewashing, or service sink used for disposing of mop water or liquid waste
✔ Dry your hands using a single-service paper towel, electric hand dryer, or clean section of continuous rolled cloth towel (if allowed in your area)	⊘ Dry your hands on your apron or a dish towel
✔ Keep fingernails trimmed and filed	⊘ Use fingernail polish or artificial fingernails without wearing gloves
✔ Vigorously rub hands to dislodge dirt	⊘ Use a dirty brush to scrub hands

Hand Antiseptics (Hand Sanitizers)

◆ Hand antiseptics may be used by food employees in addition to hand washing.

◆ Hand antiseptic lotions must never be used as a replacement for hand washing.

◆ Hand antiseptics must be formulated with safe and approved ingredients since a food employee's hand will be touching food, food-contact surfaces, or equipment and utensils after use.

Except when washing fruits and vegetables, food employees may not contact exposed, ready-to-eat foods with their bare hands. Instead, they should use suitable utensils such as deli tissue, spatulas, and tongs. Complying with the no bare hands rule is especially important when serving a highly susceptible population. Also, minimize bare hand and arm contact with exposed food that is not in ready-to-eat form.

Hand sanitizers can be used in addition to, but not in place of, hand washing.

Using Disposable Gloves

Food establishments sometimes allow their employees to use disposable gloves to help prevent contamination of foods. Gloves protect food from direct contact by human hands. Gloves must be impermeable (i.e., not allowing anything to penetrate the porous texture of the glove). Disposable gloves must be treated as a second skin. **Whatever can contaminate a human hand can also contaminate a disposable glove. Therefore, whenever hands should be washed, a new pair of disposable gloves should be worn.**

For example, if you are wearing disposable gloves and handling raw food, you must discard those gloves, wash your hands, and put on a fresh pair of gloves before you handle ready-to-eat foods. Never handle money with gloved hands unless you immediately remove and discard the gloves. Money is highly contaminated from handling. If you take latex gloves off by rolling them inside out, the inner surface of the glove is very contaminated from your skin. Again, if you take disposable gloves off, throw them away. **Never reuse or wash disposable gloves; always throw them away.**

Outer Clothing and Apparel

Work clothes and other apparel should always be clean. The appearance of a clean uniform is more appealing to your guests. During food preparation and service, it is easy for a food employee's clothing to become contaminated. If you feel you have contaminated your outer clothing, change into a new set of work clothes.

PROPER USE OF DISPOSABLE GLOVES

1. Wash Hands

2. Select Gloves

3. Put on Gloves

4. Food Handling Activity

5. Discard Gloves After Each Task

6. Wash Hands When Returning to Work

For example, if you normally wear an apron and work with raw foods, put on a fresh, clean apron before working with ready-to-eat foods. Also, never dry or wipe your hands on the apron. As soon as you do that, the apron is contaminated. Smocks and aprons help to reduce transfer of microbes to exposed food.

Keep in mind, however, that **protective apparel is similar to disposable gloves. Neither protects food when contaminated.**

Hats, hair coverings or nets, and beard restraints discourage employees from touching their hair or beard. These restraints also prevent hair from falling into food or onto food-contact surfaces. Most state and local jurisdictions require food employees to effectively keep their hair from contacting exposed food, clean equipment, utensils and linens, and unwrapped single-use articles.

According to Food Code:

The *Food Code* permits food employees to drink beverages to prevent dehydration.

◆ The beverage must be in a covered container

◆ The container must be handled in a way that prevents contamination of the employee's hands, the container, exposed food, equipment, and single-use articles.

Personal Habits

Good personal hygiene means effective health habits including bathing, washing hair, wearing clean clothing, and frequent hand washing. A food employee's fingers may be contaminated with saliva during eating and smoking. Saliva, sweat, and other body fluids can be harmful sources of contamination if they get into food. Supervisors should enforce rules against eating, chewing gum, and smoking in food preparation, service, and dishwashing areas.

Jewelry, including medical information jewelry on hands or arms, should not be worn in food production and dishwashing areas. Rings, bracelets, necklaces, earrings, watches, and other body part ornaments can harbor germs that can cause foodborne illness. Jewelry can also fall into food causing a physical hazard. A plain wedding band, that does not contain a stone, is the only type of jewelry that is permitted in food production and dishwashing areas.

Things you can do to prevent food contamination:

◆ Wear clean clothing.

◆ If your clothing is contaminated, change into a new set of work clothes.

◆ Change your apron between working with raw foods and ready-to eat foods. Aprons should be left in the work area when going on break or to the restroom.

◆ Don't dry or wipe your hands on your apron.

◆ Wear a hat, hair coverings or nets, and beard restraints to discourage you from touching your hair or beard. These restraints also prevent hair from falling into food or onto food-contact surfaces.

◆ Keep in mind, however, protective apparel is similar to a disposable glove. They no longer protect food when contaminated.

According to Food Code:

In an attempt to reduce the risk caused by sick food employees, the *Food Code* **requires employees to report to the person in charge when they have an illness and have been diagnosed by a health practitioner with:**

- Norovirus
- Hepatitis A virus
- *Shigella* spp.
- Enterohemmorrhagic or Shiga toxin-producing *Escherichia coli*
- *Salmonella* Typhi
- Symptoms of intestinal illness or flu-like symptoms (such as vomiting, diarrhea, sore throat with fever, or jaundice)
- A wound containing pus such as a boil or infected cut that is open or draining.

Personal Health

The health and hygiene of food employees are extremely important in safe food management. If a food employee is directly or indirectly exposed to *Salmonella* Typhi, *Shigella* spp., enterohemmorrhagic or Shiga toxin-producing *Escherichia coli*, Hepatitis A virus, or Norovirus, it must be reported to the supervisor, or person in charge. All these agents are easily transferred to foods, can cause illness in very low numbers, and are considered severe health hazards. The person in charge will notify the regulatory authority that a food employee is diagnosed with an illness due to any of the 5 microorganisms listed above.

If a food employee has been exposed to any of these agents, he or she may be excluded from work or be assigned to restricted activities that do not involve contact with food. Employees diagnosed with one of these diseases must not handle exposed food or have contact with clean equipment, utensils, linens, or unwrapped single-service utensils. Exclusions and restrictions for those exposed to or experiencing symptoms of a foodborne illness are specified in Appendix D of this book. When an employee is diagnosed with a reportable disease, he or she must not work until certified as safe by a licensed physician, nurse practitioner, or physician assistant.

Staphylococcus aureus is a common bacteria that is normally present in humans. About 30-50% of us carry the bacteria on our skin and in nasal passages. These bacteria are often found in infected wounds, cuts, and pimples. Infected wounds should be cleaned and completely covered by a dry, tight-fitting, impermeable bandage. Cuts or burns on a food employee's hands must be thoroughly cleaned, bandaged, and covered with a clean disposable glove. Effective hand washing is the most effective measure for limiting transfer of this bacteria to foods.

To date, there has not been a medically documented case of Acquired Immune Deficiency Syndrome (AIDS) transmitted by food. Therefore, AIDS is not considered a foodborne illness. The Americans with Disabilities Act (ADA) prohibits discrimination against people with disabilities in jobs and public accommodations. Employers may not fire or transfer individuals who have AIDS or test positive for the HIV virus away from food-handling activities. Employers must also maintain the confidentiality of employees who have AIDS or any other illness.

Employees who have been diagnosed with certain illnesses must report this to their supervisor.

Cross Contamination

Cross contamination - The transfer of microorganisms or chemicals from one food item to another.

Contaminated food contains harmful microorganisms or chemical substances/toxins that can cause foodborne illness. **The transfer of microorganisms or chemicals from one food item to another is called cross contamination.** This commonly happens when germs from raw food are transferred to a cooked or ready-to-eat food via contaminated hands, equipment, or utensils. For example, bacteria from raw chicken can be transferred to a ready-to-eat food such as lettuce or tomato when the same cutting board is used without being washed and sanitized between foods.

Cross contamination can also happen when raw foods are placed above ready-to-eat foods within cold storage facilities. Juices from the raw product can drip onto ready-to-eat foods that are stored below. This poses a health risk because ready-to-eat items will not be cooked to destroy microorganisms prior to being eaten.

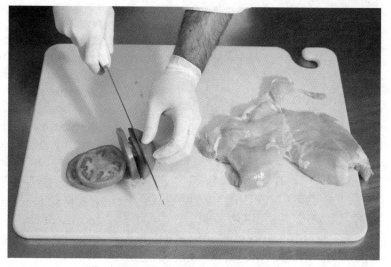

Don't mix raw and ready-to-eat foods!

In a food establishment, microorganisms and chemical substances can be transferred by a food employee, equipment and utensils, or another food.

Preventing Cross Contamination

Use color-coded cutting boards for different types of food.

(Courtesy of Food Safety Solutions)

- Always store cooked and ready-to-eat foods above raw products
- Keep raw and ready-to-eat foods separate during storage
- Use good personal hygiene and handwashing practices
- Keep all food-contact surfaces clean and sanitary
- Avoid bare hand contact with ready-to-eat foods
- Keep species of meat, poultry, and seafood separate
- Use separate equipment, such as cutting boards, for raw foods and ready-to-eat foods (color coding may be helpful for this task)
- Use clean, sanitized equipment and utensils for food production
- Prepare ready-to-eat foods first, followed by preparation of contaminated raw foods
- Prepare raw and ready-to-eat foods in separate areas.

Always keep raw foods separate from ready-to-eat foods. In cold storage, ready-to-eat foods must be stored above raw foods. Display cases, such as those used to display seafood items, should be designed to keep raw and cooked food items separate. In addition, separate buckets for in-place sanitizing solutions and wiping cloths should be used for cleaning food-contact surfaces in raw and ready-to-eat food production areas.

Other Sources of Contamination

After washing, fruits and vegetables are considered ready-to-eat foods.

Always wash fresh produce before use. Washing helps to remove soil and other contaminants. After washing, fruits and vegetables are considered ready-to-eat foods. Certain chemicals, for washing and sanitizing, may be used for raw whole fruits and vegetables. These chemicals must be nontoxic and meet the requirements set forth in the Code of Federal Regulations under title 21 CFR 173.315.

Utensils used to dispense and serve foods can also be a source of food contamination. Utensils should be properly labeled to identify the type of food they are used to dispense. During hot- or cold-holding of foods, the utensil should be stored in the food. This helps to prevent contamination from employees or customers in self-service areas. It also keeps the utensil that contains food out of the temperature danger zone. The dispensing utensils (scoops) for ice and dry bulk foods should be clean and kept in an area protected from contamination. Scoops or tongs used in customer service areas also need to be labeled and kept clean.

Animals are not permitted in food establishments unless they are being used for support or a special service (i.e., guide dogs for the blind). Food handlers must not touch animals during food preparation and service. If employees should touch an animal for any reason, they must wash their hands before returning to work.

Germs from a employee's mouth can be transferred to food when the employee uses improper tasting techniques. A food employee may not use a utensil more than once to taste food that is to be sold or served.

Wash hands or change gloves after touching live animals such as live lobsters.

Animals, rodents, and pests are common sources for food contamination. Rodents and pests usually enter food establishments during delivery or when garbage facilities are not properly maintained. An integrated pest management (IPM) program should be established and maintained in every food establishment. You will learn about IPM programs in Chapter 7, *Environmental Sanitation and Maintenance*.

Make Sure the Work Area Is Clean and Sanitary

Anything that comes in contact with food must be clean and sanitary. This includes human hands, equipment, utensils, storage areas, hot- and cold-holding areas, and self-service areas for customers. In order to protect food from contamination, effective cleaning and sanitizing procedures must be implemented and monitored. The goal of cleaning is to remove visible soil and

Keep work areas clean and sanitary.

food. The goal of sanitizing is to reduce the number of harmful microbes that may be present on a clean surface to a safe level. Chapter 6 of this book is dedicated to describing different elements of cleaning and sanitizing programs. While cleaning and sanitizing are important in all areas of a food establishment, they are especially critical in areas where ready-to-eat foods are handled and displayed.

What Do You Think?

A bad batch of potato salad

Over 318 people were reported ill from a New Jersey beach resort community after eating potato salad purchased at a local restaurant during the 4th of July holiday. Most of the people that were sick complained of intense vomiting, diarrhea, fever, nausea, abdominal cramps and severe muscle pains. The majority of people started to become ill just a few hours after eating, and most were sick from 1-2 days. Elderly victims had symptoms that lasted up to 4-5 days.

The local health department was contacted shortly after many cases of suspected food poisoning were identified at a nearby hospital emergency room. An investigation by the health department identified the likely causative agent of foodborne disease as the toxin produced by the *Staphylococcus aureus* bacteria.

Interviews with the employees who worked at the restaurant indicated that they had not had formal training in good personal hygiene practices and temperature control. The potato salad was made with cooked whole potatoes that were sliced by hand and mixed with other ingredients in a 7-gallon container and then stored in the cold room for the next day. Health department officials estimated that the cooling time, after mixing the ingredients, was over 20 hours.

How could this outbreak have been prevented?

Summary

Back to the Story...

The case study presented at the beginning of this chapter raises several areas of concern about food-handling activities employed in areas where ready-to-eat foods are prepared. There were two food-handling errors that contributed to foodborne illness. First, it is important that hands are washed well before ready-to-eat foods are prepared and served. Hand washing helps reduce the transfer of the *Staphylococcus aureus* bacteria to the food. Secondly, temperature after preparation and before/during storage was not controlled or monitored. Foods need to be rapidly cooled to 41°F (5°C) within 6 hours to control the growth of bacteria. The 20-hour cooling time in this case allowed for the growth and toxin production of the bacteria, which led to illness. Most importantly, food employees indicated that they had not been trained in safe food-handling practices. Education and training are the foundational components of effective safe food management.

Temperature abuse can occur during receiving, storing, cooking, cooling, reheating, hot-holding, and cold-holding of foods. Depending on the food type, there are specific temperature requirements to assure food safety. These requirements are discussed in the next chapter. A good rule to follow in any food establishment is, "Keep It Clean! Keep It Hot! Keep It Cold! Or Don't Keep It!"

Food employees should always use good health and hygiene practices. Clean clothing, hair restraints, and proper handwashing practices are fundamental in safe food management. Effective supervision and enforcement of proper procedures form the foundation of a successful food operation.

Control contamination with proper cleaning and sanitizing. Avoid cross contamination from one food to another. Keep foods separate and store raw foods below cooked and ready-to-eat foods. Always use proper food-handling techniques.

Keep foods at proper temperature, use good personal hygiene, and control contamination and cross contamination. These are the essentials of safe food management.

Quiz 3 (Multiple Choice)

Choose the **BEST** answer for each question.

1. Foodborne illness can be caused by:
    a. Poor personal hygiene.
    b. Cross contamination.
    c. Temperature abuse.
    d. All of the above.

2. After proper cooking, all foods that are to be held hot must be held at:
    a. 165°F (74°C) or above.
    b. 135°F (57°C) or above.
    c. 120°F (49°C) or above.
    d. Room temperature until served.

3. Sliced cooked turkey that is not kept at or below 41°F (5°C) or above 135°F (57°C) for 4 or more hours before being served is:
    a. Temperature and time abused.
    b. Cross contaminated.
    c. Safe to eat.
    d. None of the above.

4. After proper cooking, all foods that are to be held cold must be:
    a. Cooled quickly and held at 41°F (5°C) or below.
    b. Cooled quickly and held at 70°F (21°C) or above.
    c. Cooled slowly and held at 50°F (10°C) or below.
    d. Stored at room temperature until served.

5. Regarding food thermometers, which statement is **false?**
    a. They should be calibrated.
    b. They should measure temperatures between 41° F (5°C) and 135° F (57°C).
    c. They should measure temperatures between 0° F (-18°C) and 220° F (104°C).
    d. They should be approved for use in foods.

6. Which of the following methods is the most correct way to measure the temperature of frozen cuts of meat in plastic bags that have just arrived for storage?
    a. Open the package and insert probe into the center of the food.
    b. Insert the sensing probe between two packages of the frozen product.
    c. Check the ambient temperature of the truck that delivers the product.
    d. Check for ice crystals in the bottom of the bag.

7. Food employees should wash their hands after which of the following?

    a. Taking out the trash.

    b. Touching their faces.

    c. Handling raw food.

    d. All of the above.

8. A good way to prevent cross contamination of food is to:

    a. Keep raw and cooked foods separate.

    b. Properly clean and sanitize food-contact surfaces.

    c. Properly wash hands.

    d. All of the above.

9. Which of the following is an accepted personal hygiene practice?

    a. Wearing jewelry and false fingernails.

    b. Smoking and eating in food production areas.

    c. Wearing caps and hats.

    d. Wiping hands on a soiled apron.

10. Which of the following actions by a food employee would most likely cause a foodborne illness?

    a. Wearing a cap that keeps hair confined.

    b. Removing all jewelry except a plain wedding band when working.

    c. Wearing clean clothing to work each day.

    d. Not washing hands after using the toilet.

11. After chopping up raw chicken, what should be done before the cutting board is used to slice tomatoes for a salad?

    a. Rinse the board under running water and dry with a paper towel.

    b. Dry the board with a cloth and continue using to slice tomatoes.

    c. Wash, rinse, and sanitize the board before slicing the tomatoes.

    d. Turn the board over and use the other side.

12. A person who has applied for a job in your food establishment reveals that he or she is HIV positive. What must you do?

    a. Recognize that AIDS is not a disease that would be transmitted by food.

    b. Deny employment based on disease history.

    c. Call the health department and report this person was seeking work in foodservice areas.

    d. Offer a position that would require no contact with food.

Answers to the multiple-choice questions are provided in **Appendix A**.

References/Suggested Readings

Centers for Disease Control and Prevention. 2000. *Surveillance for Foodborne Disease Outbreaks—United States, 1993-1997,* (Morbidity and Mortality Weekly Report - March 17, 2000). Atlanta, GA, U.S. Department of Health and Human Services.

Centers for Disease Control and Prevention. 2006. *Surveillance for Foodborne Disease Outbreaks - United States, 2006,* (Morbidity and Mortality Weekly Report - June 12, 2009). Atlanta, GA, U.S. Department of Health and Human Services.

Code of Federal Regulations. 2001. *21 CFR 173.315 Secondary Direct Food Additives Permitted in Food for Human Consumption-Chemicals Used in Washing or to Assist in the Peeling of Fruits and Vegetables.* U.S. Government Printing Office, Washington, D.C.

Council for Agricultural Sciences and Technology. 1995. Prevention of Foodborne Illness. Dairy. *Food and Environmental Sanitation.* Vol. 15(6), 341-367.

Food and Drug Administration. 2004. *FDA Report on the Occurrence of Foodborne Illness Risk Factors in Selected Institutional Foodservice, Restaurant, and Retail Food Store Facility Types.* U.S. Public Health Service. Washington, DC. www.cfsan.fda.gov/~dms/retrsk2.html

Food and Drug Administration. 2009. *2009 Food Code.* U.S. Public Health Service, Washington, D.C.

Food Safety and Inspection Service. 1996. *Nationwide Broiler Chicken Microbiologic Baseline Data Collection Program, 1994-1995.* U.S. Department of Agriculture, Washington, D.C.

Food and Drug Administration.2009. *Employee Health and Personal Hygiene Handbook.* U.S. Public Health Service,Washington, DC. http://www.fda.gov/Food/FoodSafety/RetailFoodProtection/IndustryandRegulatoryAssistanceandTrainingResources/ucm113827.htm

Suggested Web Sites

Gateway to Government Food Safety Information

  www.foodsafety.gov

U.S. Department of Agriculture (USDA)

  www.usda.gov

Centers for Disease Control and Prevention (CDC)

  www.cdc.gov

Morbidity and Mortality Weekly Report (MMWR)

  www.cdc.gov/mmwr

Food and Drug Administration (FDA)

  www.fda.gov

USDA/FDA Food and Nutrition Information Center

  www.nal.usda.gov/fnic

Partnership for Food Safety Education

www.fightbac.org

Food Safety Day

www.foodsci.purdue.edu/publications/foodsafetyday

Hand washing for Life Institute

www.handwashingforlife.com

Clean Hands Coalition

www.cleanhandscoalition.org

Cooper-Atkins Temperature-Measuring Devices

www.cooperinstruments.com

Raytek Corporation

www.raytek.com

Brevis Corporation

www.brevis.com

Taylor Foodservice Products

www.taylorusa.com/foodsvc/food.html

DeltaTRAK, Inc.

www.deltatrak.com

Notes

Chapter

4　Following the Flow of Food

Key Terms

- ◆ Adulteration
- ◆ Aseptic packaging and processing
- ◆ Cold-holding
- ◆ Commingling
- ◆ Cooking
- ◆ Cooling
- ◆ First In, First Out (FIFO)
- ◆ Grade standards
- ◆ Hermetic packages
- ◆ Hot-holding
- ◆ Irradiation
- ◆ Modified atmosphere packaging
- ◆ Pasteurization
- ◆ Reduced oxygen packaging
- ◆ Reheat
- ◆ Sensory evaluation
- ◆ Sous vide

Food and ingredients in food establishments go through a series of steps to prepare them for sale and service. The food product flow starts with purchasing and receiving. From there, the foods and ingredients are moved into storage where they will remain until used during preparation. Finally, the finished products are displayed and/or served. It is important to understand that opportunities for contamination may exist at each step in the flow of food, and more importantly, interventions can be implemented at each step to promote the safety of the food that is provided to consumers.

Learning Points:

◆ Identify codes and symbols used to tag food products inspected by governmental agencies.

◆ Apply purchasing and receiving procedures that protect food products.

◆ Inspect equipment used to transport food products to food establishments for cleanliness and evidence of pest infestation.

◆ Use approved devices to measure temperatures in food products safely and accurately.

◆ Check for and reject defective products.

◆ Check for required product temperatures at receiving and storage.

◆ Discuss safe methods to thaw frozen foods.

◆ Identify internal temperature requirements for cooking foods.

◆ Explain the proper methods used to cool foods.

◆ Discuss the importance of employee health and hygiene related to food product flow.

◆ Employ measures to prevent contamination and cross contamination of foods.

Following the Flow of Food

The flow of food at a food establishment begins with receiving and storage. As foods are delivered they are inspected by receiving personnel and then placed quickly into storage. From storage, foods and ingredients are moved or "flow" into the preparation and handling stages of production.

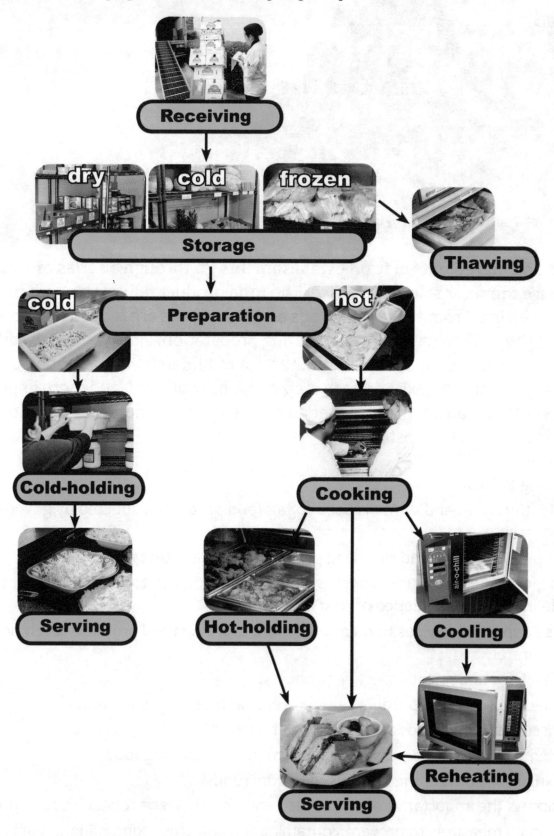

Buying from Approved Sources

Food quality and safety in food establishments begins with buying foods and ingredients from approved sources — processors and suppliers that comply with federal, state, and local food safety laws and regulations. These sources are routinely inspected to make sure they follow good agricultural and manufacturing practices. Foods prepared in a private home are not from an approved source and must not be used or sold in food establishments. Use of "home-canned" food is prohibited because of the high risk of foodborne illness, especially botulism. Always buy food from reputable suppliers that consistently follow safe food-handling practices.

Measuring Temperatures During Receiving, Storage and Preparation

As you learned in Chapter 3, maintaining safe food temperatures is an essential part of food safety management in food establishments. Properly calibrated temperature-measuring devices are used to measure the temperature of food, water, and the air inside food storage areas (refrigerators, ovens, etc.). Thermometers, thermocouples, and other devices are commonly used to measure the temperature of stored, cooked, hot-held, cold-held, and reheated foods throughout the flow of food.

Cold-holding or hot-holding equipment used for storing PHF (TCS) should be equipped with an indicating or recording thermometer to measure the temperature of the storage environment. Equipment thermometers are either built into a piece of equipment or are fastened onto shelving or other apparatus. Set them where they can be easily read. Place the sensor portion of the thermometer in the warmest part of a refrigeration unit or in the coolest part of a hot food storage unit. Some older equipment may need modification if they were designed with the sensor located in the discharge air.

Receiving

Inspecting Delivery Vehicles

The first step in the receiving process involves inspecting the vehicles that deliver food and nonfood items to your establishment. Suppliers must deliver food products to your food establishment in vehicles that are clean and in good repair. Delivery vehicles must also:

◆ Maintain perishable foods and PHF (TCS) at safe temperatures during transport

Maintaining safe product temperature is a critical part of your food safety system.

Inspect delivery vehicles during receiving.

- Be loaded in a manner that separates food items from nonfood items (detergents, household cleaners, and pesticides) to prevent contamination and cross contamination
- Protect food packages from becoming damaged and torn during transit.

When delivery vehicles arrive at the establishment, receiving personnel should inspect them. Specific items to look for during these inspections are:

- Cleanliness of the cargo area
- Temperature of refrigerated and frozen food storage areas (if applicable)
- Proper separation of food and nonfood items
- Signs of insect, rodent, or bird infestations
- Damaged packages that might result in contamination of food items.

Be prepared to reject or return products or ingredients that do not meet quality and safety standards or do not comply with applicable food laws.

Determining Food Quality

Sensory evaluation uses smell, touch, sight, and sometimes taste, to evaluate the quality of food when it is received at a food establishment. As a first step, foods should be observed for color, texture, and visual evidence of spoilage. Some common signs of spoilage are slime formation, mold growth, and discoloration.

Check foods for spoilage.

Spoiled foods frequently give off foul odors indicative of compounds such as ammonia and hydrogen sulfide (the smell of rotten eggs). These odors, caused by the breakdown of proteins through bacterial action, are usually very easy to smell.

Spoilage due to yeasts produces bubbles and an alcoholic flavor or smell. Milk develops an acidic taste and is often sour, bitter, or rancid when it spoils.

The quality and safety of a food are affected by many factors. A food that shows no signs of spoilage may not always be safe to eat. **Spoilage cannot be used as the only indicator of food safety.**

Inspect all incoming food supplies to make sure they are at the proper temperature, in sound condition, and free from filth or spoilage. Always check product containers for tears, punctures, dents, or other signs of damage. Poor receiving procedures increase the chance of:

- Theft
- Acceptance of underweight merchandise
- Contamination
- Waste
- Acceptance of products that do not meet specifications.

Signs of spoiled foods:

- Slime formation
- Mold growth
- Discoloration
- Foul odors
- Off taste.

A food may show no signs of spoilage and still be unsafe to eat. Don't use spoilage as the only indicator of food safety!

Receiving process:

Inspect all incoming food supplies. Check:

- Product temperatures
- Condition of product - no damage such as rips, tears, punctures, or dents
- That products are free from filth and spoilage.

Food establishments should schedule deliveries for off-peak times and have enough staff and space on hand to receive products quickly and correctly. Move incoming shipments to storage as soon as they arrive. Merchandise that is damaged, spoiled, or otherwise unfit for sale or use must be properly disposed of, held by the establishment for credit, or returned to the distributor. Distressed merchandise must be put aside to prevent contamination of other foods, equipment, utensils, linens, or single-service or single-use articles.

Merchandise that is damaged, spoiled, or otherwise unfit for sale should be rejected by receiving clerks. Store rejected foods away from good foods.

Store rejected foods away from good foods.

Whatever the size of the establishment, receiving requires:

◆ Prompt handling
◆ Quality control procedures
◆ Trained staff who know:
  ◆ Product specifications
  ◆ Coding
  ◆ Proper checking of product temperatures
  ◆ Proper handling of rejected merchandise.

Packaged Foods

Foods come in many different types of packages, including cans, bottles, jars, pouches, tubs, trays, bags, and boxes. The common purpose of the package is to:

◆ Protect the contents from contamination
◆ Provide a source of information about its nutritional contents
◆ Provide advertising material
◆ Make the product more convenient for customers to transport, prepare, and serve.

Dry foods such as flour, sugar, rice, and beans are commonly packaged in bags. These are not PHF (TCS) and can be safely stored at room temperature. Check packages for damage and signs of contamination from chemicals and other substances that could cause foodborne illness.

Reduced oxygen packaging - special packaging techniques designed to reduce the amount of oxygen in foods to prevent spoilage.

Reduced Oxygen Packaging (ROP)

For many foods, oxygen in the air can increase chemical breakdown (oxidation) and microbial spoilage. Many food processors use **reduced oxygen packaging** to help overcome the effects of oxygen and preserve foods. Some examples of reduced oxygen packaging include vacuum packaging, modified atmosphere packaging, and sous vide foods.

Benefits to ROP packaging

- It reduces oxygen to prevent the growth of aerobic bacteria, yeast, and molds largely responsible for the off odors, slime, texture changes, and other forms of spoilage
- It prevents chemical reactions that can produce off odors and color changes in foods
- It reduces product shrinkage by preventing water loss.

Common ROP packaging choices

- **Vacuum packaging** — removes air from a package and hermetically seals the package so a near-perfect vacuum remains inside.
- **Modified atmosphere packaging (MAP)** — a process that employs a gas flushing and sealing process or reduction of oxygen through respiration of vegetables or microbial action.
- **Sous vide** — a process where fresh raw foods are sealed in a plastic pouch and the air is removed by vacuum. The pouch is cooked at a low temperature and rapidly cooled to 38°F (3°C) or below or frozen.
- **Cook-chill** — a process that uses a plastic bag filled with hot cooked food from which air has been forced out and which is closed with a plastic or metal crimp.
- **Controlled atmosphere packaging (CAP)** — a system that maintains the desired atmosphere within a package throughout the shelf life of a product by the use of agents to bind or scavenge oxygen or a small packet containing compounds to emit a gas.

KEY TERM

Hermetic packages - a container that is sealed to prevent the entrance of bacteria, molds, yeasts, and filth, as long as it remains intact.

Vacuum Packaging - Vacuum packaging removes air from a package and hermetically seals the package so a near-perfect vacuum remains inside. The term hermetic refers to a container that has been sealed to prevent the entrance of bacteria, molds, yeasts, and filth, as long as it remains intact. The most commonly used **hermetic packages** are metal cans, glass jars, and flexible plastic packages. These containers will also stop the entry of bacteria, yeasts, molds and other types of contamination as long as they remain undamaged. Upon receiving, metal cans must be checked for defects.

Checking for defective cans

Leaky or bulging

◆ Do not accept cans if they leak or bulge at either end
◆ Swollen ends on a can indicate gas is being produced inside
◆ Gas may be caused by a chemical reaction between the food and the metal in the container or by the bacteria and other microbes inside the can.

Dented

◆ Dents in cans do not harm the contents unless they have actually penetrated the can or the seam
◆ Dents found in the side seams or end seams of a can are the most important
◆ Do not accept cans if damage to these areas can affect the physical integrity of the can and may allow microorganisms to enter through tiny pinhole leaks
◆ Shipments with many dented cans or torn labels indicate poor handling and storage procedures by the supplier.

Rusty

◆ Rust does not harm contents unless it has penetrated the can or seam
◆ Rusty cans indicate exposure to excess moisture.

Modified atmosphere packaging (MAP) - The MAP process helps preserve foods by replacing some or all of the oxygen inside the package with other gases such as carbon dioxide or nitrogen. MAP is used with a wide range of products including meat, fish, pre-cut lettuce, baked products, cheese, coffee, nuts, potato chips, and dried fruit. The MAP process has been successful in extending the shelf life of many products, and it reduces the amount of additives and preservatives required to prevent deterioration of the food. However, since low oxygen environments can create conditions conducive to the growth of anaerobic bacteria like *Clostridium botulinum* (see Chapter 2), proper handling of MAP foods is essential. MAP technology requires adequate refrigeration to be maintained during

KEY TERM

Modified atmosphere packaging (MAP) - a process of preserving foods by replacing some or all of the oxygen inside the package with other gases such as carbon dioxide or nitrogen.

the entire shelf life of PHF (TCS). During the receiving process, employees must inspect PHF (TCS) products in modified atmosphere packages to make sure the package is in sound condition and the food is at 41°F (5°C) or below when it arrives at the establishment.

Sous vide - The term **sous vide** is French for "without air." It is a specialized packaging process where fresh raw foods are sealed in plastic pouches and the air is removed by vacuum. The pouches are then cooked at a low temperature and rapidly cooled to 38°F (3°C) or below or frozen. The low cooking temperature of the sous vide process kills spoilage microorganisms. However, it does not destroy *C. botulinum* bacteria and their spores. Therefore, PHF (TCS) items processed using sous vide technology must be kept out of the food temperature danger zone during transport, storage, and display. Receiving personnel must inspect sous vide products to make sure the package is in sound condition and the food is at 41°F (5°C) or below as it arrives at the establishment. Some sous vide foods may require even colder temperatures during receipt and cold storage. Check with the manufacturer about cold-holding temperatures for any sous vide foods you purchase. If extended shelf life is desired, a temperature of 38°F (3°C) or lower must be maintained at all times. After receiving, the products should be moved quickly into refrigerated storage until ready for use.

Regulations regarding ROP - Some products cannot be packed by ROP unless the food establishment is approved for the activity and inspected by the

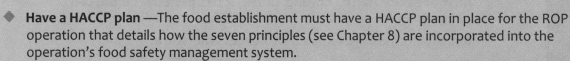

According to Food Code:

The *Food Code* requirements for food establishments that use ROP technology:

Food Code
U.S. Public Health Service
FDA
2009

◆ **Have a HACCP plan** —The food establishment must have a HACCP plan in place for the ROP operation that details how the seven principles (see Chapter 8) are incorporated into the operation's food safety management system.

◆ **Use two microbial growth barriers** —Because *Clostridium botulinum* and *Listeria monocytogenes* are significant health hazards, the establishment is required to have two microbial growth barriers. Time/temperature are commonly used as one barrier, but they need to be coupled with the use of pH, A_w, or product formulation to assure *Clostridium botulinum* and *Listeria monocytogenes* will not grow inside the package.

◆ **Maintain foods at proper temperature** —All PHF (TCS) in ROP that rely on refrigeration as a barrier to microbial growth must be maintained at 41°F (5°C) or below.

◆ **Set shelf life** —The refrigerated shelf life is to be no more than 14 days from packaging to consumption or the original manufacturer's "sell-by" or "use-by" date, whichever comes first.

◆ **Proper label warnings** —Packages must be prominently and conspicuously labeled on the principal display panel with the instructions to maintain the food at 41°F (5°C) or below and to discard refrigerated food if within 14 days of its packaging it is not served for on-premises consumption or consumed if served or sold for off-premises consumption.

◆ **Use-by or sell-by dates** —Each container must bear a use-by or sell-by date. This date cannot exceed 14 days from packaging or repackaging to consumption, or the original manufacturer's "sell-by" or "use-by" date, whichever occurs first.

◆ **Employee training** —Employees responsible for the ROP process must receive training that will enable them to understand the key components of the process, the equipment used, and the specific procedures that must be followed to ensure critical limits identified in the HACCP plan have been met. The training program must also outline the employee's responsibilities for monitoring and documenting the process and describe what corrective actions they must take when critical limits are not met.

appropriate regulatory authority. These products include raw and smoked fish, soft cheeses (ricotta, cottage cheese, and cheese spreads), and combinations of cheese and other ingredients such as vegetables, meat, or fish. Contact your local regulatory authority to obtain a complete set of guidelines for ROP foods and to seek a variance for all food manufacturing/processing operations based on the prior approval of a HACCP plan.

Irradiation

Food **irradiation** is a preservation technique used by some food processing industries. This process involves exposing food to controlled amounts of radiation in order to destroy disease-causing microorganisms and delay spoilage. This will increase the safety and extend the shelf life of the food.

The acceptance of irradiated foods has been slow due to consumers' concerns about the safety of foods preserved in this manner. Contrary to many myths, irradiated food is not radioactive and does not pose a risk to the health and safety of people who eat it. Foods processed with irradiation are just as nutritious and flavorful as other foods that have been cooked, canned, or frozen.

Federal law requires irradiated food to be labeled with the international symbol for irradiation called a "radura." This symbol must be accompanied by the words "Treated with Irradiation" or "Treated with Radiation."

Irradiation of food can effectively reduce or eliminate pathogens and spoilage microbes while maintaining the quality of most foods. This is a technology proven to be safe and should be welcomed by customers as an effective food preservation technique.

Red Meat Products

Most red meat and meat products sold in the United States come from cattle (beef), calves (veal), hogs (ham, pork, and bacon), sheep (mutton), and young sheep (lamb). These products are inspected for wholesomeness by officials of the U.S. Department of Agriculture (USDA) or state agencies. Animals must be inspected for wholesomeness to make certain they are free of disease and not adulterated. **Adulterated** food contains filth or is otherwise decomposed and unfit for human consumption. The USDA also offers voluntary meat grading services. **Grade standards** for meat represent the culinary quality or palatability of the meat and are not measures of product safety. The figure below contains examples of inspections and grading stamps applied to products approved for use.

Inspection   **Grade**

USDA inspection and grade stamps for beef, veal, and lamb

> The FDA has approved food irradiation for a variety of foods including fruits, vegetables, grains, spices, poultry, pork, lamb, and ground meat.

KEY TERMS

Irradiation - the process of exposing food to controlled amounts of radiation in order to destroy disease-causing microorganisms and delay spoilage.

Adulterated - food containing filth or is otherwise decomposed and unfit for human consumption

Grade standards - standards of meat quality based on culinary quality or palatability. Not a measurement of product safety.

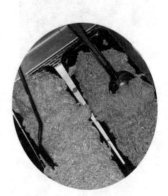

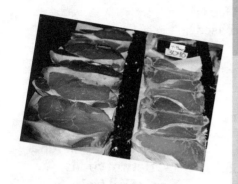

Receiving red meat products		
	Accept	**Reject**
Fresh	◆ Temperature below 41°F (5°C) at delivery ◆ Firm and elastic to the touch ◆ Has characteristic aromas.	◆ Product temperature exceeds 41°F (5°C) at delivery. ◆ Off odors ◆ Sliminess.
Frozen	◆ Solidly frozen ◆ Packaged to prevent freezer burn.	◆ Look for signs of freezing and thawing and refreezing such as frozen blood juices in the bottom of the container or the presence of large ice crystals on the surface of the product.

Meat and meat products are available in several forms such as fresh, frozen, cured, smoked, dried, and canned. Since raw meats are PHF (TCS), never accept them if there is any sign of contamination, temperature abuse, or spoilage.

Move fresh meat into refrigerated storage as quickly as possible. Frozen products should be moved from the delivery truck to the freezer while they are still solidly frozen.

Poultry

Some common examples of poultry are chicken, turkey, duck, and geese. USDA or state inspectors must inspect all poultry products to make certain they are wholesome and not adulterated. Inspected poultry products carry a USDA seal on the individual package or on bulk cartons. Usually poultry is graded also for quality. Grade A poultry must have good overall shape and appearance, be meaty, be practically free from defects, and have a well-developed layer of fat in the skin.

Inspection

Grade

USDA inspection and grade stamps for poultry

Poultry products support the growth of disease-causing and spoilage microorganisms. The intestinal tract and skin of poultry may contain a variety of foodborne disease bacteria, including *Salmonella* spp. and *Campylobacter jejuni*. The near neutral pH, high moisture, and high protein content of poultry make it an ideal material for bacteria to grow in and on.

Receiving poultry products		
	✔ Accept	**⊘ Reject**
Fresh	◆ Temperature below 41°F (5°C) at delivery ◆ Firm and elastic to the touch ◆ Well-developed layer of fat in the skin ◆ Has characteristic aromas.	◆ Product temperature exceeds 41°F (5°C) at delivery ◆ Off odors ◆ Sliminess ◆ Stickiness under the wings ◆ Discolored or darkened wing tips.
Frozen	◆ Solidly frozen ◆ Packaged to prevent freezer burn.	◆ Look for signs of freezing and thawing and refreezing such as frozen blood juices in the bottom of the container or the presence of large ice crystals on the surface of the product.

> **Inspect poultry very carefully... bacteria grows easily on poultry due to its:**
>
> ◆ Near neutral pH
> ◆ High moisture
> ◆ High protein content.

Poultry products are also vulnerable to spoilage caused by enzymes and spoilage bacteria. Poultry should be packaged on a bed of ice that drains away from the meat as it melts and held at or below 41°F (5°C).

Game Animals

Game animals are not permitted for sale in food establishments unless they meet federal code regulations. This ban does not apply to commercially raised game animals approved by regulatory agencies, field-dressed game allowed by state codes, or exotic species of animals that must meet the same standards as those of other game animals.

Game animals commercially raised for food must be raised, slaughtered, and processed according to standards used for meat and poultry. Common examples of animals raised away from the wild and used for food are farm-raised buffalo, ostrich, and alligator. The USDA inspects the slaughter and processing of this meat in the usual manner.

Eggs

Most food establishments sell and use eggs in one form or another. Eggs are usually purchased by federal grades, the most common being AA, A, and B. Grades for eggs are based on exterior and interior conditions of the egg.

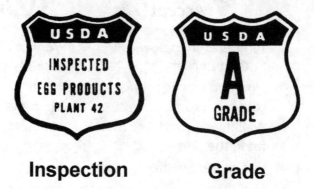

Inspection   **Grade**

USDA inspection and grade stamps for eggs

Salmonella enteritidis bacteria enter the inside of the egg as it is formed inside the hen. The eggshell surface may contain *Salmonella* spp. bacteria, especially if the shell is soiled with chicken droppings. Even if the shell is not cracked, bacteria can enter through the pores in the egg's shell.

	Receiving eggs and egg products	
	✔ **Accept**	⊘ **Reject**
Raw shell eggs	◆ Clean ◆ Fresh ◆ Free of cracks or checks ◆ Refrigerated at an air temperature of 45°F (7°C) or below when delivered.	◆ Dirty ◆ Cracked ◆ Off odors ◆ Temperature over 45°F (7°C).
Liquid egg products	◆ In a sealed container ◆ Kept at 41°F (5°C).	◆ Temperature over 41°F (5°C).

Food Code
U.S. Public Health Service
FDA
2009

According to Food Code:

Egg products:

◆ As a safeguard against *Salmonella* spp., the FDA requires all egg products, such as liquid, frozen, and dry eggs, to be pasteurized to render them *Salmonella* free.

◆ Pasteurized egg products should be in a sealed container and kept at 41°F (5°C).

◆ Egg containers should carry labels that verify the contents have been pasteurized.

Raw shell eggs should be clean, fresh, free of cracks or checks, and refrigerated at an air temperature of 45°F (7°C) or below when delivered. Shell eggs that have not been treated to destroy all viable *Salmonella* shall be stored and displayed in refrigerated equipment that maintains an ambient temperature of 45°F (7°C) or less. These eggs must be labeled to include safe handling instructions. The egg, when opened, should have no noticeable odor, a firm yolk, and the white should cling to the yolk. Reject eggs that are dirty or cracked, and remember washing eggs only increases the possibility of contamination.

An egg product is defined as an egg without its shell. These products must be pasteurized to render them *Salmonella* free. They are well suited for facilities such as day care centers, hospitals, and nursing homes that offer food to people in highly susceptible populations.

Fluid and Dry Milk and Milk Products

This food group includes fluid and dry milk, cheese, butter, ice cream, and other types of milk products. When receiving milk and milk products, make certain they have been pasteurized. **Pasteurization** destroys all disease-causing microorganisms in the milk and reduces the total number of bacteria, thus increasing shelf life. All market milk must be Grade A quality. Pasteurization also destroys natural milk enzymes that might shorten the shelf life of the products.

Aseptic processing and packaging involves placing a sterilized product into a sterilized package that is then sealed under sterile conditions. For example, ultra-high temperature pasteurized milk (commonly labeled as UHT milk) and its package are sterilized separately and then combined and sealed in a sterile environment. Aseptic processing and packaging allows the food to retain more color, texture, taste, and nutrition than it would if subjected to more heat-intensive food preservation methods such as canning and bottling. UHT products can be stored safely for several weeks if kept under refrigeration. These products can also be stored without refrigeration for short periods of time (Longrèe and Armbruster).

Aseptic packaging of individual creamers

Fluid milk

Receiving fluid and dry milk and milk products		
	✔ **Accept**	⊘ **Reject**
Fluid milk	◆ Refrigerated at an air temperature of 45°F (7°C) or below when delivered.	◆ Past expiration date. ◆ Temperature above 45°F (7°C).
Cheese	◆ Received at 41°F (5°C) ◆ Proper color ◆ Proper flavor.	◆ Temperature over 41°F (5°C) ◆ Moldy ◆ Damaged packaging.
Butter	◆ Received at 41°F (5°C) ◆ Firm texture ◆ Uniform color.	◆ Rancid odor ◆ Moldy ◆ Damaged packaging.

Fluid Milk

Under the Pasteurized Milk Ordinance, fluid milk should be received at 45°F (7°C) or less. It should be refrigerated immediately upon delivery and cooled within four hours to 41°F (5°C) or below. Individual containers of milk should be clearly marked with an expiration date and the name of the dairy plant that produced it. Check the expiration date of all dairy products before using them.

Cheese

Cheese should be received at 41°F (5°C) and checked for the proper color, flavor, and characteristics. Reject the product if it contains mold that is not a normal part of the cheese or if the rind or package is damaged.

Butter

Butter is made from pasteurized cream. Since disease-causing and spoilage bacteria and mold may grow in butter, handle the product as a perishable item. The most common type of deterioration in butter is the development of a strong rancid odor and flavor. Ensure butter has a firm texture, even color, and is free of mold. Packaged butter should be received intact at 41°F (5°C) and provide protection for the contents.

Fish

Fish includes finfish harvested from saltwater and freshwater and seafood that comes mainly from saltwater. Seafood consists of molluscan shellfish and crustaceans. Molluscan shellfish include oysters, clams, mussels, and scallops. Crustaceans include shrimp, lobster, and crab. Oysters, shrimp, catfish, salmon, and a few other types of finfish and seafood are being raised on fish farms using a technique called aquaculture.

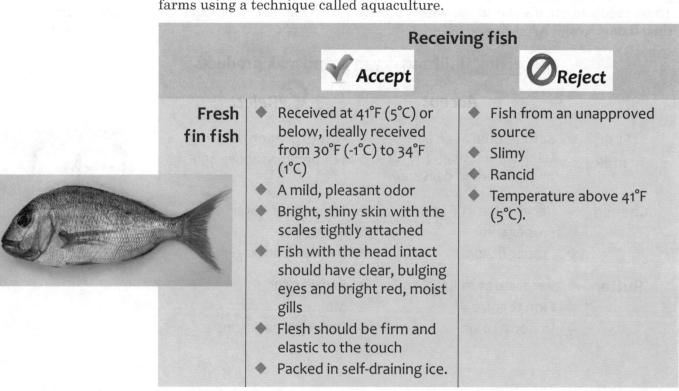

	Receiving fish	
	✔ Accept	⊘ Reject
Fresh fin fish	◆ Received at 41°F (5°C) or below, ideally received from 30°F (-1°C) to 34°F (1°C) ◆ A mild, pleasant odor ◆ Bright, shiny skin with the scales tightly attached ◆ Fish with the head intact should have clear, bulging eyes and bright red, moist gills ◆ Flesh should be firm and elastic to the touch ◆ Packed in self-draining ice.	◆ Fish from an unapproved source ◆ Slimy ◆ Rancid ◆ Temperature above 41°F (5°C).

Fish should be received at 41°F (5°C) or below and shellfish may be received at 45°F (7°C) or below. For better quality and shelf life, the optimum fish and shellfish receiving temperature often ranges from 30°F (-1°C) to 34°F (1°C). The quality of fish and seafood is measured by smell and appearance. These products are generally more perishable than red meats, even when stored in a refrigerator or freezer. They are commonly packed in self-draining ice to prevent drying and to maximize the shelf life of the food. The slime covering the outside of fish contains a variety of bacteria that makes them highly susceptible to contamination and microbial spoilage. Fish are also rich in unsaturated fatty acids that are susceptible to oxidation and the development of off flavors and rancidity.

Fish must be commercially and legally caught or harvested — except when caught recreationally — and approved for sale by the regulatory authority. All fish suppliers and warehouse operations must comply with the seafood HACCP program as required in the Code of Federal Regulations (21 CFR 123). Ready-to-eat raw, raw-marinated, partially cooked, or marinated partially cooked fish must be frozen to time and temperature guidelines that meet *Food Code* specifications in order to kill parasites. Records must be retained that show how the product was handled. These requirements do not apply to molluscan shellfish, tuna species specified in the *Food Code*, and fish species, such as salmon, that are raised using aquaculture techniques.

Shellfish must be purchased from sources approved by the FDA and the health departments of states located along the coastline where the shellfish is harvested. Shellfish transported from one state to another must come from sources listed in the Interstate Certified Shellfish Shippers List. The reason for requiring tight control over molluscan shellfish is to reduce the risk of infectious Hepatitis A virus and other foodborne illnesses that may result from eating raw or insufficiently cooked forms of this product. Molluscan shellfish caught recreationally may not be used or sold in food establishments.

Ready-to-eat raw, raw-marinated, partially cooked, or marinated-partially cooked fish must be frozen to time and temperature guidelines that meet *Food Code* specifications in order to kill parasites.

Purchase molluscan shellfish from approved sources.

Receiving shellfish		
	✔ **Accept**	⊘ **Reject**
Shellfish	◆ Received at 45°F (7°C) or below ◆ Clean ◆ Contains source identification tags.	◆ Shellstock from unapproved sources ◆ Contains dead shellfish ◆ Broken shells ◆ Missing source identification tags ◆ Temperature above 41°F (5°C).

When received at a food establishment, molluscan shellfish should be reasonably free of mud, dead shellfish, and shellfish with broken shells. Damaged shellfish must be discarded.

According to Food Code:

Molluscan shellfish must be purchased in containers that bear legible source identification tags or labels fastened to the container by the harvester and each dealer that shucks, ships, or reships the shellstock. Molluscan shellfish tags must contain the following information:

◆ The harvester's identification number

◆ The date of harvesting

◆ An identification of the harvest location or aquaculture site including an abbreviation of the state or country in which the shellfish are harvested

◆ The shellfish type and quantity

◆ A statement in bold, capitalized type that says, "**THIS TAG IS REQUIRED TO BE ATTACHED UNTIL CONTAINER IS EMPTY OR IS RETAGGED AND THEREAFTER KEPT ON FILE FOR 90 DAYS FROM THE DATE THE SHELLSTOCK CONTAINER IS EMPTIED.**"

> **Shellstock containers must be kept until every piece of product from that batch is gone.**

KEY TERM

Commingling - combining shellfish harvested on different days or from different growing areas as identified on the tag or label, or combining shellfish from containers with different container codes or different shucking dates.

If seafood is suspected of being the source of foodborne illness, the investigating team can use the tags to determine where and when the product was harvested and processed.

For display purposes, shellstock may be removed from the tagged or labeled container in which they are received. The shellstock must be placed on drained ice or held in a display container. Care must be taken to protect shellstock from contamination during display.

The identity of the source of shellstock that has been removed from a tagged or labeled container for display must be preserved. This can be accomplished by using an approved record-keeping system that keeps shellstock tags or labels in sequence based upon the date when, or dates during which, the shellstock are sold or served. Shellstock from one tagged or labeled container must not be commingled with shellstock from another container before being ordered by the customer. **Commingling** is combining shellfish harvested on different days or from different growing areas as identified on the tag or label, or combining shellfish from containers with different container codes or different shucking dates.

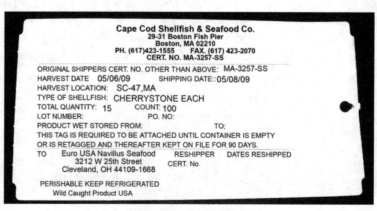

Example of shellfish tag

Fruits and Vegetables

Most fruits and vegetables have a short shelf life. They continue to ripen even after they are picked. Therefore, they may become over ripe if not properly handled. Microorganisms found in water and soil can also cause fruits and vegetables to spoil. Fruits and vegetables hold their top quality for only a few days.

Purchase raw fruits and vegetables from approved sources and wash them thoroughly to remove soil and other contaminants before they are cut, combined with other ingredients, cooked, served, or offered for human consumption in a ready-to-eat form.

Though not required, whole raw fruits and vegetables may be washed using cleaners. The FDA reminds us that pathogens can be transferred from one tomato to another when batches of whole tomatoes are soaked in standing water. Therefore, the agency recommends that whole tomatoes be washed under running water. An anti-microbial rinse can be used to reduce the number of microorganisms present on the surface. When these types of chemicals are used, they must meet the requirements in the Code of Federal Regulations (21 CFR 173.315). The fruits and vegetables should also be rinsed to remove as much of the residues of these chemicals as possible.

Some products, like wild mushrooms, may only be used if they have been inspected and approved by a mushroom-identification expert who is approved by the regulatory authority. Beware of fresh mushrooms packaged in Styrofoam trays and covered with plastic shrink-wrap. Mushrooms use up the oxygen inside the package quickly. Unless holes are poked in the plastic wrap that covers the package to permit oxygen inside, oxygen-free conditions may occur that are favorable for the growth of *C. botulinum* bacteria.

According to Food Code:

◆ The FDA now considers sprouts, cut melons, cut tomatoes, and cut leafy greens to be PHF (TCS) and requires them to be handled in the same manner as other PHF (TCS).

DID YOU KNOW?

Most fresh fruits are not considered to be PHF (TCS) products. Their acidity and/or outer skin prevent the entry and growth of disease-causing bacteria. Nonetheless, the number of cases of foodborne illness linked to produce has increased significantly in recent years. This is largely due to increased consumption of fresh fruits and vegetables, the global supply of these products, and the emergence of microbes that can cause disease with a low number of organisms.

Norovirus, *Salmonella* spp., Shiga toxin-producing *E. coli*, *Shigella* spp., Hepatitis A virus, and *Cyclospora* spp. can be infective with only a few cells. Therefore, they do not require a PHF (TCS) to multiply and cause foodborne illness.

Juice

Juice includes the liquid extracted from one or more fruits or vegetables, purees of the edible portions of one or more fruits or vegetables, or any concentrates of such liquid or puree. According to the *Food Code*, this group of foods includes juice as a whole beverage, an ingredient of a beverage, and a puree as an ingredient of a beverage.

Most of the juices sold in food establishments are obtained from processors in a prepackaged form. These juices must be obtained from a processor that has a HACCP system in place. In most instances, the processor will pasteurize or otherwise treat the juice to attain 99.999% reduction of the most resistant disease-causing microorganisms of public health significance.

Juice packaged in a food establishment must:

◆ Be treated under a HACCP plan as specified in 8-201.13 and 8-201.14 of the *Food Code* to attain a 99.999% reduction of the most resistant microorganisms of public health significance, or if the juice is not treated to destroy pathogens.

◆ Bear a warning label that informs customers "This has not been pasteurized and, therefore, may contain harmful bacteria that can cause serious illness in children, the elderly, and persons with weakened immune systems."

WARNING: This product has not been pasteurized and, therefore, may contain harmful bacteria that can cause serious illness in children, the elderly and persons with weakened immune systems.

Juice warning label

Purees of fruits or vegetables that are not used as beverages or ingredients in beverages are not required to comply with the HACCP requirements.

Frozen Foods

Frozen products must be solidly frozen when delivered. Check the temperature of frozen foods by placing the sensing portion of a thermometer between two packages. Receiving personnel should also look for signs the product has been thawed and refrozen.

Common signs of thawing and refreezing are:

◆ Large ice crystals or frost on the surface of the food
◆ Frozen liquid or juice at the bottom of the package
◆ Mushy soft products.

Check frozen foods for signs of thawing and refreezing.

Reject frozen foods that are not solidly frozen or show signs of temperature abuse.

Storage of Food

KEY TERMS

First-in, first-out (FIFO) - method of stock rotation to assure older foods are used first.

Employees must check incoming shipments carefully and quickly move received items to proper storage. Stock rotation is a very important part of effective food storage. A **first-in, first-out (FIFO)** method of stock rotation helps assure older foods are used first. Product containers should be marked with a date or other readily identifiable code to help employees know which product has been in storage longest. When expecting food shipments, always make certain the older stock is moved to the front of the storage area to make room for the newly arriving products.

Date coding facilitates proper stock rotation.

Types of Storage

The three most common types of food storage areas in food establishments are the:

◆ Refrigerator
◆ Freezer
◆ Dry storage.

Refrigerated Storage

Refrigerated storage is used to maintain the safety of PHF (TCS) and to maintain the quality of other perishable foods. It slows down microbial growth and controls quality by holding foods at 41°F (5°C) or below. Some common types of refrigerated storage equipment found in food establishments are:

◆ Walk-in
◆ Reach-in
◆ Under-the-counter
◆ Upright
◆ Cold display units, among others.

Reach-in cooler

In order to maintain the temperature of PHF (TCS) at 41°F (5°C) or below, equipment should maintain the air temperature in the storage compartment at about 38°F (3° C). Fish and shellfish are especially vulnerable to spoilage and should be stored at colder temperatures ranging from 30°F to 34°F (-1° to 1°C). Some fruits and vegetables, such as bananas and potatoes, undergo undesirable chemical changes when they are refrigerated. Therefore, although fruits and vegetables are perishable products, not all types should be refrigerated. For perishability, fresh fruits and vegetables requiring refrigeration should be stored at temperatures between 33°F (1°C) and 41°F (5°C). Unpasteurized juices and PHF (TCS) fruits and vegetables, like cut melons, cut tomatoes, cut leafy greens, and sprouts must be stored at 41°F (5°C) or below.

Cold storage units should maintain air temperature of 38°F (3°C) or below. Check temperatures daily in the warmest part of the unit to ensure accuracy using an indicating or recording thermometer. Keep cooler doors closed as much as possible to maintain proper temperature.

Store raw products under cooked or ready-to-eat foods.

Important procedures for cold storage

◆ Keep refrigerated foods at 41°F (5°C) or below and frozen foods solidly frozen during storage.

◆ Rotate refrigerated and frozen foods on a FIFO basis and store foods in covered containers that are properly labeled and dated.

◆ Store foods in refrigerated and freezer storage areas at least 6 inches off the floor and space products to allow the cold air to circulate around them.

◆ Store raw products under cooked or ready-to-eat foods to prevent cross contamination.

◆ Keep different species of raw animal foods separate during storage. If limited storage space makes it necessary to store different species in the same area of the refrigerator, store poultry on the bottom shelf; ground meat and pork on the middle shelf; and fish, eggs, and other cuts of red meat on the top shelf.

Freezer Storage

Freezer storage is designed to keep foods solidly frozen. Freezer equipment must also be equipped with indicating or recording thermometers to monitor the temperature of the ambient air inside the unit. If your freezer is not frost free, defrost it regularly to ensure proper operation. Wrap frozen foods and transfer them to the refrigerated storage area until the defrosting process is complete.

Although bacteria are generally not destroyed by freezing, parasites can be killed if foods are frozen at the proper temperature for the proper length of time. Guidelines have been established for destroying parasites in raw marinated and marinated, partially cooked fish.

The *Food Code* allows yellowfin, bigeye, bluefin northern, bluefin southern, and certain other species of tuna to be served or sold in a raw, raw-marinated, or partially cooked ready-to-eat form without freezing.

Guidelines for destroying parasites by freezing

◆ Food should be frozen throughout to -4°F (-20°C) and held for 7 days in a freezer, or;

◆ Food should be frozen throughout to -31°F (-35°C) using a blast chiller, and held at that temperature for 15 hours, or;

◆ Food should be frozen throughout to -31°F (-35°C) and stored at -4°F (-20°C) for at least 24 hours.

Sushimi

Dry Storage

Products in dry storage areas are usually packed in labeled cans, bottles, jars, and bags. The area should have a room temperature of 50°F (10°C) to 70°F (21°C) with a relative humidity of 50 to 60% to maximize shelf life of stored products. Windows should be blocked or shaded.

◆ Use slatted shelves that allow circulation of air, are at least 6 inches off the floor, and are away from the wall. This allows for cleaning under the shelving and discourages pest harborage.

◆ When bulk items are moved into bulk food grade containers with tight-fitting lids, include codes, labels and dates.

◆ Scoops and other utensils should be food grade and have long handles that keep hands from touching food.

◆ Do not use toilet rooms, locker areas, mechanical rooms, and similar spaces for storage of food, single-service items, paper goods, or equipment and utensils.

◆ Do not expose products to overhead water and sewer lines unless the lines are shielded to interfere with potential drips.

Store dry foods on slatted shelving.

Chemical Storage

Toxic chemicals, such as cleaners, sanitizers, and pesticides, are commonly used and sold in food establishments. Most of these products can be poisonous if consumed accidentally. Many chemicals used in food establishments are poisonous if consumed. Others can cause irritation to skin and respiratory system.

Guidelines for chemical storage

◆ All products must be labeled and kept separate from food products. If an adequate storage area is not available, use a locked cabinet to store the chemicals.

◆ Identify the chemical and include directions on proper use.

◆ Train employees on how to use these products safely.

◆ It is a good practice to post lists of instructions so users can easily see when and how to use the products.

A good label identifies the chemical, provides directions on how to use it safely, and instructs people on what first-aid measures to use in case of accidents.

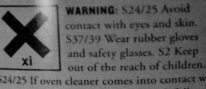

Store chemicals away from food.

Storage Conditions for Foods

The following chart shows ideal storage conditions for various foods.

Food Type	Storage conditions
Meat and meat products 	◆ Store for up to 3 weeks at temperatures between 28°F (-2°C) and 32°F (0°C) and a relative humidity between 85 and 90% ◆ Cold temperatures extend the shelf life of red meats by slowing down the growth of bacteria that cause spoilage and reduce shrink loss ◆ Store for several months when held at 0°F (-18°C) or below ◆ Frozen meats must be wrapped in moisture-proof paper to prevent them from drying out ◆ Packaging for frozen foods should also be strong, flexible, and protect against light ◆ Use by manufacturer's shelf-life criteria.
Poultry 	◆ Store at temperatures between 28°F (-2°C) and 32°F (0°C) for short periods of time ◆ A relative humidity of 75 to 85% is recommended, as excessive humidity causes sliminess due to excessive bacterial growth ◆ Poultry should be wrapped carefully to prevent dehydration, contamination, and loss of quality ◆ Frozen poultry and poultry products can be stored for 4 to 6 months when held at 0°F (-18°C) or below ◆ Use by manufacturer's shelf-life criteria.
Whole shell eggs 	◆ Keep fresh for up to 2 weeks when stored at 41°F (5°C) or below ◆ It is recommended to store eggs at 34°F (1°C) to 38°F (3°C) to maintain optimum quality ◆ Keep eggs covered and store them away from onions and other foods that have a strong odor ◆ Discard eggs that are dirty or cracked ◆ Always make sure to wash your hands after handling whole shell eggs ◆ Use by manufacturer's shelf-life criteria.
Egg products (such as egg whites and yolks)	◆ Are pasteurized to destroy *Salmonella* bacteria ◆ Store products at 41°F (5°C) or below ◆ Store frozen eggs at 0°F (-18°C) or below and keep them frozen until time for defrosting ◆ Once dried eggs have been reconstituted, they are considered potentially hazardous and must be stored at 41°F (5°C) or below ◆ Use by manufacturer's shelf-life criteria.

(cont.)

Food Type	Storage conditions	(continued)

Milk

- ◆ Pasteurized milk may be held at 41°F (5°C) or less for up to 10 days or longer.
- ◆ The optimal storage temperature for fluid milk is 34°F (1°C) to 38°F (3°C), and the shelf life of milk is shortened significantly at higher storage temperatures.
- ◆ Milk also picks up odors from other foods. Store milk in an area away from onions and other foods that give off odors.
- ◆ Use by manufacturer's shelf-life criteria.

Fish and shellfish

- ◆ More perishable than red meats even when refrigerated or frozen.
- ◆ Fish and shellfish should be stored at temperatures ranging from 30°F (-1C) to 34°F (1°C).
- ◆ Fish should be kept on crushed ice drained away from the product or solidly frozen. Recommend using fresh fish within 24 hours or less.
- ◆ Shellfish shells should close when tapped. Dead shellfish must be discarded. Lobsters and clams should be kept alive until cooked or frozen. Keep shellstock tags for 90 days after the container is emptied. If a foodborne outbreak occurs, the tags help identify the source.

Fresh fruits and vegetables

- ◆ Many fresh fruits and vegetables are refrigerated for quality.
- ◆ Cut melons, leafy greens and tomatoes are PHF (TCS) and must be stored at 41°F (5°C) or below.
- ◆ If fruits and vegetables arrive packed in airtight film, notify your supplier to correct this issue to allow the product to respire.
- ◆ Produce should not be washed before storage — wash before using.
- ◆ Prevent cross contamination by segregating fresh produce, including tomatoes, from other foods in refrigeration units.
- ◆ Proper circulation is necessary to maintain freshness and firmness. Discard fruits that begin to spoil.
- ◆ Whole citrus fruits and bananas should not be refrigerated.

Reduced oxygen packaging (ROP) foods

- ◆ ROP products are perishable foods and must be kept at temperatures recommended by the processor.
- ◆ Most will need refrigeration at 41°F (5°C) or below; if frozen, keep solidly frozen until thawed and used.
- ◆ Check expiration dates before using; discard out-of-date products.
- ◆ Do not use packages that have signs of microbial growth (slime, bubbles, molds, etc.).

Preparation and Service

The preparation and service of foods can involve one or more steps. Small food establishments, such as convenience stores, buy foods in ready-to-eat forms which are stored until sold. Large operations, such as restaurants, supermarkets, and institutional kitchens, prepare large quantities of food. Preparation and service are complex operations in these larger establishments. They can involve many steps and span several hours or days.

Regardless of how many steps are involved in food production and service, foodborne illness prevention requires effective food safety measures that ensure good personal hygiene and avoid cross contamination and temperature abuse.

During preparation, an important technique to follow is "small batch" preparation. Food preparation is usually done at room temperature. This is several degrees into the temperature danger zone. Therefore, you must limit the amount of time the food is in the temperature danger zone by working with small and manageable amounts of PHF (TCS) and ingredients.

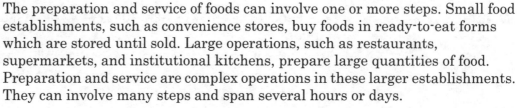

Prepare food in small batches to reduce the amount of time the food is in the temperature danger zone.

Ingredient Substitution

For meal solutions and home meal replacements, there is normally a recipe for the products prepared in the food establishment. The recipe usually includes a list of ingredients and instructions for how to prepare, store, and label the food item.

◆ When one or more of the original ingredients is not available for a recipe, other ingredients may be substituted so the food item can still be prepared and sold

◆ All ingredient substitutions should be identified in the recipe before preparation

◆ Ingredient substitutions must never compromise the safety of the food and should not be allowed unless they are identified and allowed in the recipe.

Avoiding Temperature Abuse

Monitor temperatures throughout preparation and service.

Temperature and time abuse is when food is kept in the temperature danger zone, 41°F (5°C) to 135°F (57°C), long enough for harmful organisms to grow.

◆ Monitoring and controlling food temperatures are extremely effective ways to minimize the risks of foodborne illnesses.

◆ Thermometers are used for stored, cooked, hot-held, cold-held, and reheated foods. Before using a thermometer, make sure it is clean, sanitary, and properly calibrated. Always insert the "sensor" portion or probe stem of the thermometer into the thickest part of the food. In most instances, this will be at the center of the food product or container.

Thawing

The *Food Code* requires raw animal foods to be thawed in less than 4 hours including the time it takes for preparation for cooking or to lower the food temperature to 41°F (5°C) under refrigeration. Thawed portions of ready-to-eat foods should not be allowed to rise above 41°F (5°C) when using the cool water thawing process.

Food establishments will sometimes use a slacking (defrosting) process to moderate the temperature of foods prior to cooking or reheating. During the slacking process, foods can be defrosted under refrigeration that maintains the food at 41°F (5°C) or less or at any temperature if the food remains frozen. The slacking process is typically used with previously block-frozen food such as spinach.

Foods can also be thawed in a microwave oven. When this process is used, the food must be cooked immediately after thawing. Food can be cooked in the microwave oven or it can be transferred to conventional cooking equipment such as a stove or oven. Foods thawed in a microwave oven may not be stored in a refrigerator for preparation at a later time.

Under no circumstances should PHF (TCS) products be thawed or slacked at room temperature. Room temperature thawing puts foods in the temperature danger zone—the very thing you don't want to have happen. When foods are thawed at room temperature, the outer surface of the food thaws first and will soon reach room temperature. Microbial growth occurs very quickly at room temperature.

Thawing individual steaks

Guidelines for thawing PHF (TCS)

Refrigeration

◆ Use refrigeration that maintains the food temperature at 41°F (5°C) or below.

Submerge under Cool, Running Water

Completely submerge under running water:

◆ At a water temperature of 70°F (21°C) or below

◆ With enough water force to remove contaminants from the surface of the food

◆ For a period of time that does not allow thawed portions of ready-to-eat foods to rise above 41°F (5°C)

◆ For a period of time that does not allow thawed portions of a raw animal food requiring cooking to be in the temperature danger zone for more than a total time of 4 hours.

As Part of the Cooking Process

◆ Some foods can be thawed and cooked in one continuous process, (i.e., frozen pizza).

Thawing for Immediate Service

◆ Use any procedure (i.e., microwave oven) that thaws a portion of frozen ready-to-eat food prepared for immediate service in response to an individual customer's order.

Cold Storage

Most harmful microorganisms start to grow at temperatures above 41°F (5°C). Some bacteria, such as *Listeria monocytogenes*, can grow slowly at temperatures below 41°F (5°C). Proper **cold-holding** is required to control the growth of disease-causing microbes and extend the shelf life of perishable products.

KEY TERM

Cold-holding - refers to the safe temperature range [less than 41°F (5°C)] for maintaining foods prior to service for consumption.

Cold-holding at a glance: refrigerators, display case and cold service bars

Cold, raw PHF (TCS) (like meat and poultry)
Temperature: Below 41°F (5°C)

◆ Physical barriers should be in place to separate different species (meat, poultry, seafood) and to separate raw from ready-to-eat foods.

Fish and seafood
Temperature: Below 41°F (5°C)

◆ Use ice from potable water
◆ Transport ice in "approved food-contact" containers
◆ Liquid must be drained from ice to prevent contamination
◆ Ice in contact with fish and shellfish is considered contaminated —DO NOT REUSE
◆ Cooked and raw products should be kept separate.

Cold, ready-to-eat PHF (TCS)
Temperature: Below 41°F (5°C) up to 7 calendar days

◆ If held more than 24 hours, the *Food Code* recommends prepared and held products be marked to indicate the date or day by which the food shall be consumed on the premises, sold, or discarded.
◆ The day of preparation or the day the original container is opened in the food establishment is counted as Day 1.
◆ The day or date marked by the food establishment may not exceed a manufacturer's use-by date if the manufacturer determined the use-by date based on food safety.

Ready-to-eat salads
Temperature: Below 41°F (5°C)

◆ Pre-chill ingredients before using
◆ Prepare small batches to assure food is not in the temperature danger zone too long.

Note: If the target cold-holding temperature for PHF (TCS) is 41°F (5°C) or below, it will be necessary to maintain the ambient temperature in the coolers at approximately 38°F (3°C) or below.

Frozen, Ready-to-Eat Foods

When large amounts of food are removed from the freezer, they should be marked to indicate the date by which the food must be sold. Ready-to-eat PHF (TCS) must be used within 7 calendar days or less after the food is removed from the freezer, minus the time before freezing. Subtract any time if the food is maintained at 41°F (5°C) or less before freezing.

When displaying ready-to-eat PHF (TCS) like prepared salads and luncheon meats, be sure the refrigerator unit can maintain a safe cold-holding temperature. This may be more difficult in open top and open front refrigerated display cases that do not have doors. Refrigerated display cases have a "safe load line." This line indicates the level below which foods must be stored to ensure the food is held at the proper temperature. It is also important to store foods so the discharge or return air vents are not blocked.

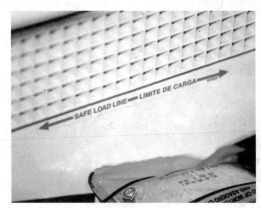

Safe load line

Prepackaged Foods

A refrigerated, ready-to-eat, PHF (TCS) prepared and held in a food establishment for more than 24 hours, or originating from an original container prepared and packaged by a food processing plant and opened at the food establishment must be discarded if it:

◆ Exceeds the prescribed time and temperature requirements, or;

◆ Is in a container or package that does not bear a date or day, or;

◆ Is marked with a date or day that exceeds the time and temperature combination described above.

Cooking

Meat, poultry, fish, seafood, eggs, and unpasteurized milk should not be prepared and served raw or rare. Establishments that choose to serve raw foods increase the risk of causing a foodborne illness. Raw animal foods need to be cooked to the proper temperatures to be safe. Many states and jurisdictions require an advisory be posted to warn consumers of the risk.

Check temperatures during cooking.

Disease-causing organisms are killed by exposing them to lethal temperatures for a prescribed amount of time. Therefore, cooking requirements in retail food operations must include a final cook temperature and the amount of time the food is to be held at that temperature in order to destroy the pathogens.

Cooking - the act of providing sufficient heat and time to a given food to effect a change in the food and destroy foodborne pathogens inherent in that food.

Food establishments commonly use stoves, conventional ovens, and microwave ovens when **cooking** foods. Because heat transfer can be different depending on the heating source, final temperature requirements have been set for conventional oven cooking and microwave cooking.

Cooking guidelines for PHF (TCS)

Food Type	Minimum Internal Temperature	Minimum Time Held at Internal Temperature Before Serving
Meat roast (rare)	130°F (54°C) 140°F (60°C)	112 min. 12 min.
Eggs, meat and pork (other than roasts), fish	145°F (63°C)	15 sec.
Ground meat, mechanically tenderized meat, ground pork, and ground game animals	155°F (68°C)	15 sec.
Meat roast (medium), pork roast, and ham	145°F (63°C)	4 min.
All poultry, ground poultry, stuffed meats, and stuffed food products	165°F (74°C)	15 sec.

Note: When microwave cooking, heat raw animal foods to a temperature of 165°F (74°C) in all parts of the food.

Serving undercooked hamburgers and other ground meats is not an option for items on a children's menu

The internal temperature of raw animal foods cooked in a microwave oven must reach 165°F (74°C) or above.

Casseroles and other foods that contain a combination of raw ingredients such as meat and poultry must be cooked to a final temperature that coincides with the highest risk food. In this case, 165°F (74°C) is required to destroy pathogens that may be found in the poultry. Food mixtures, such as chili and beef stew, must be cooked to 165°F (74°C) to assure proper destruction of disease-causing agents.

When cooking foods in the microwave oven, the distribution of heat is often uneven. Stirring and rotating the food during the cooking process will enable heat to be distributed more evenly. The *Food Code* requires raw animal foods cooked in a microwave oven to be heated to 165°F (74°C) in all parts of the food. As a common practice, foods cooked in a microwave oven should be allowed to stand covered for two minutes before serving to allow the heat inside the product to disperse more evenly.

For most foods, the internal temperature will be measured by inserting the probe of a thermometer or thermocouple into the center or thickest part of the food mass. This will give you an accurate reading of the internal temperature of the product.

Cooling

Foods are in the temperature danger zone during **cooling** and there is no way to avoid it. After proper cooking, PHF (TCS) need to be cooled from 135°F (57°C) to 41°F (5°C) as rapidly as possible. The *Food Code* recommends hot foods, not used for immediate service or hot display, be cooled from 135°F (57°C) to 70°F (21°C) within 2 hours, and from 135°F (57°C) to 41°F (5°C) or less within 6 hours.

PHF (TCS) prepared from ingredients such as reconstituted foods and canned tuna, which have been held at room temperature, must be cooled to 41°F (5°C) or less within 4 hours.

According to Food Code:
Using a noncontinuous cooking process

The *Food Code* permits food establishments to use a noncontinuous cooking process. In noncontinuous cooking, the initial heating of food is intentionally halted so it may be cooled and held for complete cooking at a later time prior to sale or service. Raw animal foods that are cooked using a noncontinuous cooking process shall be:

◆ Heated for no more than 60 minutes in the initial heating process
◆ Cooled immediately after heating using the proper procedures for PHF (TCS)
◆ Held frozen or cold as required for PHF (TCS)
◆ Cooked using a process that heats all parts of the food to a temperature of at least 165°F (74°C) for 15 seconds prior to sale or service
◆ Cooled according to the time and temperature parameters for cooked PHF (TCS) if not either held hot, served immediately, or held using time as a public health control.

Ice bath

Place food in shallow pans 2 to 3 inches deep.

Safe cooling methods

Improper cooling is one of the leading contributors to foodborne illness in retail food establishments

Common methods for reducing cooling time

◆ Blast chillers
◆ Walk-in coolers, loosely covered
◆ Use containers that facilitate heat transfer (stainless steel)
◆ Transfer food into shallow pans that will allow for a product depth of 3 inches or less
◆ Transfer food into smaller containers
◆ Place container of hot food in an ice water bath
◆ Stir food while cooling
◆ Use cooling paddles to stir the food
◆ Add ice as an ingredient directly to a condensed food.

Foods must pass through the temperature danger zone as quickly as possible. Never assume any one method is working without checking the temperature and time foods take to cool. Always depend on the thermometer reading with any of the methods you use to cool food. The following chart shows the differences in cooling time based on cooling method and portion size.

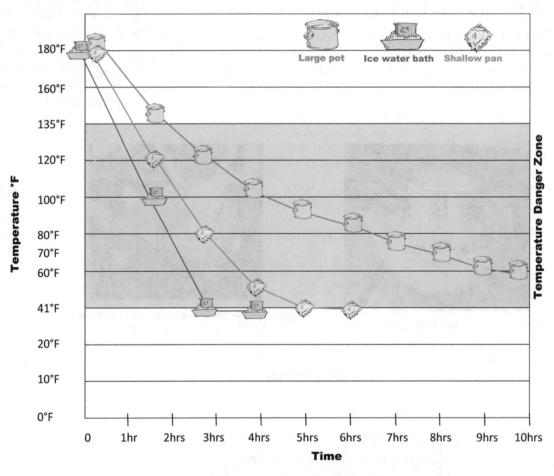

Different cooling methods for chili

Hot-holding and Reheating

All PHF (TCS) that have been cooked and are intended to be held hot (not cooled, stored, and reheated) must be maintained at 135°F (57°C) or above.

◆ **Hot-holding** is also required when hot PHF (TCS) are delivered to sites away from the food establishment

◆ During hot-holding, never add fresh product to existing product and always work in small batches

◆ **Reheat** to at least 165°F (74°C) within 2 hours.

Ready-to-eat foods commercially prepared and packaged by a food processing plant under regulatory inspection should be free of harmful microorganisms. Therefore, these foods may be reheated to a temperature of at least 135°F (57°C) for hot-holding [rather than a minimum of 165°F (74°C)]. Commercially prepared and cooked soups would be a good example of a food that would only need to be reheated to 135°F (57°C) or above followed by hot-holding. Meat roasts cooked to the proper temperature and time may be held at 130°F (54°C) or above.

While temperature is usually the most important factor in controlling microbes, there are some situations where controlling time can also be used. That is why there is an allowance of four hours in the temperature danger zone for ready-to-eat PHF (TCS) held for food service or immediate consumption. Sliced pizza and fried chicken are good examples. If these PHF (TCS) were hot-held at 135°F (57°C) or above, the food may dry out and the quality may deteriorate very quickly.

Hot-holding—Store hot-held foods at 135°F (57°C) or above.

According to Food Code:

Food Code
U.S. Public Health Service
FDA
2009

The *Food Code* allows ready-to-eat PHF (TCS) to be stored <u>without</u> temperature control when the following guidelines are used:

◆ Up to 4 hours, after which it must be consumed immediately or discarded

◆ Up to 6 hours for refrigerated food, if the food temperature is at 41°F (5°C) or below when initially removed from temperature control, and as long as the food temperature never exceeds 70°F (21°C).

Time as a Public Health Control

The *Food Code* allows ready-to-eat PHF (TCS) held for sale or service using time only as the public health control to be sold or served at any temperature upon a consumer's request. For example, slices of pizza, chicken wings, and egg rolls may be heated for taste during the holding time before being served. These foods are commonly served at short-term buffets or catered meals.

Serving Safe Food

Do   Don't

Employees must practice good personal hygiene when serving food. This starts with a clean uniform and an effective hair restraint.

◆ Food handlers should avoid touching food with their bare hands

◆ They can use tongs, serving spoons, disposable gloves, or deli tissue when handling meats, cheeses, prepared salads, or when making sandwiches

◆ Employees must hold serving utensils by the handle only, and they must never touch the part of the utensil that comes into contact with food

◆ A single utensil should be used for each food item, and the utensil should be stored in the food between uses

◆ Always store serving utensils in a way that permits the employee to grab the handle without touching the food.

Handle eating utensils properly

Food that has been served or sold to and is in the possession of a customer may not be returned and offered for service or sale to another customer. Two acceptable exceptions to this rule are:

◆ A container of a non-PHF (TCS) (i.e., a narrow-neck bottle of catsup or steak sauce) that is dispensed in a way that protects the food from contamination and the container is closed between uses, or;

◆ Non-PHF (TCS) such as crackers, salt, or pepper in an unopened, original package and that is maintained in a sound condition.

Employees must remember to wash their hands after touching soiled equipment, utensils, and cloths. If disposable gloves are used, a fresh pair of gloves should be put on immediately prior to handling any food products.

Discarding or Reconditioning Food

A food that is unsafe, adulterated, or not honestly presented shall be reworked or reconditioned using a procedure approved by the regulatory authority in the jurisdiction, or it must be discarded.

Food must be discarded if it is not from an approved source or has been contaminated by food employees, consumers, or other persons via soiled hands, bodily discharges (i.e., coughs and sneezes), or other means.

Ready-to-eat foods must be discarded if they have been contaminated by an employee who has been restricted or excluded as described in Appendix D of this book.

Refilling Returnable Containers

A take-home food container returned to a food establishment may not be refilled with a PHF (TCS) at the establishment. A food-specific container for beverages may be refilled at a food establishment if:

◆ Only a beverage that is not a PHF (TCS) is dispensed into the container

◆ The design of the container and of the rinsing equipment and the nature of the beverage, when considered together, allow effective cleaning of the container at home or in the food establishment

Coffee cups can be refilled.

- Facilities for rinsing before refilling returned containers with fresh, hot water that is under pressure and not recirculated are provided as part of the dispensing system
- The consumer-owned container returned to the food establishment for refilling is refilled for sale or service only to the same consumer, and;
- The container is refilled by an employee of the food establishment or the owner of the container if the beverage system includes a contamination-free transfer process that cannot be bypassed by the container owner
- Personal takeout beverage containers, such as thermally insulated bottles, non spill coffee cups, and promotional beverage glasses, may be refilled by employees or the consumer if the refilling process will protect the food and food-contact surface of the container from contamination.

Self-Service Bar

Self-service salad and hot food buffet bars are very popular in food establishments. They offer convenience and a wide range of selections for customers. The most important food safety goals for this type of operation are to:

- Protect foods from contamination by customers
- Keep foods out of the temperature danger zone.

Rules for Self-Service Bars:

- A properly installed sneeze guard protects the food from contamination by your customers.
- Never place raw animal foods on self-service bars, except for ready-to-eat foods like sushi and shellfish or meats that will be cooked on the premises. Keep hot-held PHF (TCS) at 135°F (57°C) or above and cold foods at 41°F (5°C) or below.
- Use clean and sanitized utensils in a self-service bar and replace any utensils that become contaminated or soiled.
- Use only one utensil for each food item and store it in the food between uses.

If customers are allowed to visit the self-service bar more than once, they must be given a clean plate or bowl for each trip. This will reduce the risk of contaminating food on display at the bar. Beverage cups and glasses may be reused to get refills.

Temporary Facilities and Mobile Food Facilities

A temporary food establishment (TFE) is defined by the *Food Code* as a food establishment that operates for a period of no more than 14 consecutive days in conjunction with a single event or celebration. TFEs may operate either indoors or outdoors and often have limited physical and sanitary facilities available.

Some food establishments are offering food by way of temporary and mobile facilities. A variety of foods can be prepared and served from temporary stands and trailers. Some examples include sandwiches, pizza, barbecue ribs, gyros, corn on the cob, confections (i.e., cotton candy, funnel cakes, or elephant ears), and beverages.

Mobile facilities are trucks and trailers used to cater events located away from the establishment. The extent of food items offered through catering operations is practically unlimited.

Mobile food facility

The same food safety practices employed in other areas of the establishment must also be applied at temporary, mobile facilities, and in-store product demonstrations. In particular, food must be protected from:

Temporary food facility

◆ Temperature abuse
◆ Infected employees who practice poor personal hygiene and use improper food-handling practices
◆ Contamination and cross contamination.

Employees shall use tongs, utensils, and deli tissue to avoid bare hand contact with food. Disposable gloves can provide an additional barrier against contamination. However, gloves must not be viewed as a substitute for proper hand washing.

Food must be protected from contamination and cross contamination. Food on display must be protected from contamination by customers and employees. Food equipment must be designed and constructed to make it smooth, easily cleanable, nontoxic and nonabsorbent.

> **Employees must not eat and smoke around food, and they must wash their hands whenever they are contaminated.**

Use disposable utensils whenever possible. When it is necessary to use multiple-use utensils, they must be washed and sanitized using a four-step process:

1. Wash in hot, soapy water.
2. Rinse in clean water.
3. Use chemical sanitizing rinse.
4. Air-dry.

Garbage and paper wastes should be placed in containers lined with plastic bags and equipped with tight-fitting lids. Food and wastes must be kept covered to avoid attracting insects, rodents, and other pests.

Vending Machines

A vending machine is a self-service device which dispenses individually sized servings of food and beverages after a customer inserts a coin, paper currency, token, card, key, or makes a payment by another means. Some vending machines dispense food and beverages in bulk while others dispense products in individually wrapped packages. Vending machines are available that will dispense PHF (TCS), ready-to-eat foods (i.e., sandwiches, french fries, dairy products, and soup) and non-PHF (TCS) items (i.e., candy, snack chips, pastries, coffee, and soft drinks).

Vending machines that store and dispense PHF (TCS) must have adequate refrigeration and/or heating units, insulation, and controls to keep cold foods cold and hot foods hot. These machines must also be equipped with an automatic control that prevents the machine from vending food if there is a power failure, mechanical failure or other condition that prevents cold food from being maintained at 41°F (5°C) or below and hot food maintained at 135°F (57°C) or above.

The temperature cutoff requirement does not apply:

◆ In a cold food machine during a period not to exceed 30 minutes immediately after the machine is filled, serviced or restocked, or;

◆ In a hot food machine during a period not to exceed 120 minutes immediately after the machine is filled, serviced or restocked.

Refrigerated, ready-to-eat, PHF (TCS) prepared in a food establishment and dispensed through a vending machine (with an automatic shutoff control) shall be discarded if it:

◆ Exceeds the time and temperature combinations prescribed by the *Food Code,* or;

◆ Is not correctly date labeled.

Home Meal Replacement

Home meal replacements and meal solutions are the terms most often used to refer to high-quality meals prepared away from home but eaten at home. These types of products are now a multi-billion dollar business for food establishments throughout the country.

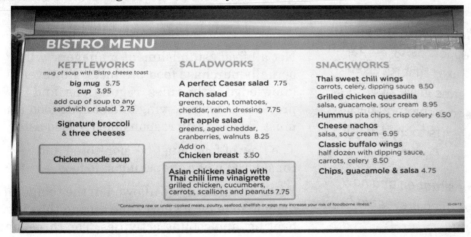

Home meal replacements come in "ready-to-cook," "ready-to-heat," and "ready-to-eat" varieties.

Ready-to-eat foods are a common variety of home meal replacement.

The most common types of home meal replacements include "ready-to-cook," "ready-to-heat," and "ready-to-eat" products. All three varieties are designed to save time and effort for families that are too tired to cook at the end of the day. Ready-to-eat foods are the most common variety of home meal replacements. It is a good idea to establish "sell-by" or "best if used by" dates and codes for these foods.

Home meal replacements should be labeled so customers understand how to keep the product safe when they take it home. Warn them against keeping the food in the car or keeping it at room temperature when they get home. Pamphlets and brochures stuffed into bags or stapled to the front of bags can help educate customers about safe handling of home meal replacements. Operators must be able to show they've done everything in their power, including written documentation and following industry standards, to make the food safe.

What Do You Think?

Salmonella infections associated with raw produce items

The Centers for Disease Control and Prevention (CDC), the U.S. Food and Drug Administration (FDA) and state/local regulatory officials investigated a multi-state outbreak of *Salmonella* serotype Saint Paul in 2008. According to information released by the CDC, 1,442 people from 43 states became ill during the outbreak.

Initially, investigators thought the outbreak was linked to consumption of raw tomatoes including red plums, red Roma, round red tomatoes, and products containing these tomatoes. However, it was later learned that contaminated jalapeño peppers, serrano peppers, and cilantro were also potential sources of the *Salmonella*. The FDA traced the source of the jalapeño peppers associated with illness to distributors in Texas that received jalapeño peppers from farms in Mexico. The farms provided produce to a common packing facility in Mexico that exports to the United States.

What went wrong in this situation and how could this outbreak have been prevented?

Summary

Back to the Story... Raw produce was identified as the source of the *Salmonella* infections described in the case study at the beginning of the chapter. Foodborne disease outbreak investigation results support the conclusion that jalapeño peppers were a major vehicle by which the *Salmonella* bacteria were transmitted and serrano peppers also were a vehicle. Raw red tomatoes may have also been a vehicle, particularly early in the outbreak.

No specific cause of contamination was identified for this outbreak. However, investigators suspect the produce items may have been contaminated at the farm or during processing or distribution.

Purchase raw fruits and vegetables from approved sources and wash them thoroughly under running water to remove soil, pesticide residues and other contaminants. Bacterial growth can also be slowed by storing produce out of the temperature danger zone.

Not all food establishment managers will actually purchase food products. However, knowledge of the rules, regulations, and procedures for receiving and storing food is a must for everyone responsible for food safety.

The flow of food must be monitored and considered when preparing safe food.

Foods should only be allowed in the temperature danger zone for a short time during thawing, heating, and cooling activities. The three main contributors to foodborne illness in food establishments are:

◆ Temperature and time abuse
◆ Cross contamination
◆ Poor personal health and hygiene practices by food handlers.

These factors must be controlled throughout the flow of food to assure food safety. Preparation and service are critical processes in your establishment because they are the last steps you take before your customer eats the food.

Quiz 4 (Multiple Choice)

Please choose the BEST answer to the questions.

1. Which of the following is (not) a rule that should be followed by food establishments when purchasing foods?

    a. Food prepared in a private home may not be used or offered for sale for human consumption.
    b. Buyers should only purchase food that is safe, wholesome, and from an approved source.
    c. Avoid the use of commercially raised game animals as sources of meat and poultry.
    d. Only buy meat and poultry that has been inspected for wholesomeness by the U.S. Department of Agriculture.

2. The most accurate way to measure the temperature of packaged frozen foods at receiving is by:

    a. Inserting the sensing probe of a temperature-measuring device into the center of the food until the recorded temperature stabilizes.
    b. Inserting the sensing probe between two packages of frozen foods until the recorded temperature stabilizes.
    c. Measuring the ambient temperature of the freezer unit on the delivery vehicle.
    d. Looking for signs of freezing and thawing, such as large ice crystals on the surface of the food and frozen juices on the bottom of the container.

3. Which of the following foods should **not** be rejected upon delivery at a food establishment?

    a. Fresh fish that has dull, sunken eyes and soft flesh.
    b. Poultry with darkened wing tips and soft flesh.
    c. Canned fruit with a small amount of rust on the lid of the can.
    d. Fresh meat products that are delivered at 45°F (7°C).

4. Which of the following statements about fish and shellfish is **false**?

    a. Fish and shellfish are less likely to spoil than red meat and poultry.
    b. Quality in fish and shellfish can be measured by smell and appearance.
    c. Most types of fish that are to be eaten raw must be commercially frozen prior to consumption.
    d. Shellfish must be purchased from sources approved by the FDA and the health departments of states located along the coastline where the shellfish is harvested.

5. Which of the following storage practices should prompt a manager to take corrective action?

    a. Raw poultry is stored above pasta salad in the walk-in cooler.
    b. Food in the walk-in cooler is stored on slatted shelves that are 6 inches above the floor.
    c. Products in the dry storage area are being rotated on a FIFO basis.
    d. Cleaners, sanitizers, and pesticides are stored in a room across the hall from the dry storage area.

6. Fresh shell eggs should arrive in a transport container that has an ambient temperature of:

    a. $38°F$ ($3°C$)
    b. $45°F$ ($7°C$).
    c. $55°F$ ($13°C$).
    d. $70°F$ ($21°C$).

7. Shellstock tags must be kept on file in a food establishment for at least
    ____ days from the date a container of molluscan shellfish is emptied.

    a. 15 days.
    b. 30 days.
    c. 60 days.
    d. 90 days.

8. The safest way to thaw a PHF (TCS) is:

    a. Under an infrared light.
    b. At room temperature.
    c. In the refrigerator.
    d. Under cool running water at 70°F (21°C).

9. Hot foods must be held at ____ or above and cold foods should be held at
    ____ or below when on display in a food establishment.

    a. 165°F (74°C); 41°F (5°C).
    b. 165°F (74°C); 32°F (0°C).
    c. 135°F (57°C); 41°F (5°C).
    d. 135°F (57°C); 32°F (0°C).

10. Poultry, ground poultry, and stuffed meats should be cooked to an internal
    temperature of _____ for 15 seconds to be considered safe.

    a. 135°F (57°C).
    b. 145°F (63°C).
    c. 155°F (68°C).
    d. 165°F (74°C).

11. Ground meat, ground pork, mechanically tenderized meat, and ground
    game animals must be cooked to an internal temperature of
    _____ for 15 seconds to be considered safe.

    a. 135°F (57°C).
    b. 145°F (63°C).
    c. 155°F (68°C).
    d. 165°F (74°C).

12. Raw animal foods cooked in a microwave oven must be heated to an
    internal temperature of _____ for 15 seconds to be considered
    safe.

    a. 135°F (57°C).
    b. 145°F (63°C).
    c. 155°F (68°C).
    d. 165°F (74°C).

13. All PHF (TCS) that have been cooked and cooled must be reheated to an internal temperature of _____ within 2 hours to be considered safe.

    a. 135°F (57°C).

    b. 145°F (63°C).

    c. 155°F (68°C).

    d. 165°F (74°C).

14. According to the *Food* Code, foods that are cooked and then cooled must be cooled in which of the following ways?

    a. From 135°F (57°C) to 41°F (5°C) in 12 hours.

    b. From 135°F (57°C) to 41°F (5°C) in 8 hours.

    c. From 135°F (57°C) to 70°F (21°C) in 2 hours, and from 135°F (57°C) to 41°F (5°C) within 6 hours.

    d. From 135°F (57°C) to 70°F (21°C) in 4 hours, and from 70°F (21°C) to 41°F (5°C) within 2 hours.

15. Which of the following statements about cold storage is **false**?

    a. Refrigerated and frozen foods must be stored at least 6 inches off the floor in a walk-in refrigerator/freezer.

    b. Refrigerated and frozen foods do not have to be rotated on a FIFO basis.

    c. Raw products should be stored under cooked and ready-to-eat foods to prevent cross contamination.

    d. When possible, different species of raw animal foods should be kept separate during storage.

Answers to the multiple-choice questions are provided in **Appendix A**.

References/Suggested Readings

Food and Drug Administration. 2009. *2009 Food Code*. U.S. Public Health Service, Washington, D.C.

Food and Drug Administration. 2001. *Fish and Fishery Products Hazards and Control Guide, 3rd ed.* U.S. Public Health Service, Washington, D.C.

Longrèe, Karla, and G. Armbruster. 1996. *Quantity Food Sanitation*. John Wiley and Sons, Inc., New York, NY.

Thayer, David W., et al.. 1996. *Radiation Pasteurization of Food. Council for Agricultural Science and Technology Issue Paper, No. 7,* April, 1996.

Suggested Web Sites

Gateway to Government Food Safety Information

  www.foodsafety.gov

U.S. Department of Agriculture

  www.usda.gov

Centers for Disease Control and Prevention

  www.cdc.gov

Food and Drug Administration

  www.fda.gov

The American Egg Board
  www.aeb.org
American Meat Institute
  www.meatami.org
The National Chicken Council
  www.eatchicken.com
U.S. Poultry and Egg Association
  www.poultryegg.org
USDA Food Safety and Inspection Service (FSIS_
  www.fsis.usda.gov
USDA/FDA Food and Nutrition Information Center
  www.nal.usda.gov/fnic
Environmental Protection Agency
  www.epa.gov
The Food Marketing Institute
  www.fmi.org
Conference for Food Protection
  www.foodprotect.org
Food Irradiation Processing Alliance
  www.fipa.us
International Food Information Council
  www.ific.org
National Frozen and Refrigerated Foods Association, Inc.
  http://www.nfraweb.org
Partnerships for Food Safety Education
  www.fightbac.org
Produce Marketing Association
  www.pma.com
United Fresh Produce Association
  www.unitedfresh.org

Notes

Chapter 5

Facilities, Equipment, and Utensils

Key Terms

- Easy-to-clean
- Equipment
- Food-contact surfaces
- Kitchenware
  - Single-use articles
  - Tableware
  - Utensils
- Non-food-contact surfaces
- Sealed
- Smooth

The design of your facility and the equipment and utensils used, contribute to a safe and sanitary environment for food preparation and service. Your facility should promote the smooth flow of work, provide adequate facilities and space for performing work efficiently, and contribute to high sanitation standards.

Learning Points:

◆ Recognize the importance of properly maintaining equipment and utensils and the influence it has on food safety.

◆ Describe how work tasks are conducted in work centers.

◆ Recognize how the preparation and service of food flows through food establishments.

◆ Remember the basic design requirements that apply to floor- and counter-mounted equipment used in food establishments.

◆ Recognize the different types of cooking, refrigeration, preparation and dishwashing equipment available for use in food establishments.

◆ Describe how proper installation and maintenance affect the operation of equipment used during the course of food production, holding, display, and handling.

◆ Explain the role of proper lighting in food production and warewashing areas.

◆ Explain how proper heating, air conditioning, and ventilation affect food sanitation and employee comfort and productivity in food establishments.

Design, Layout, and Facilities

The design, layout, and facilities provided in a food establishment must be consistent with the types of foods being prepared and sold there. The equipment used will be determined by the menu and types of preparation procedures required to produce food items. A layout that works well in one establishment may not necessarily be suitable for another that produces and sells different food items. The design and layout of storage, production, display, and warewashing areas should provide an environment in which work may be conducted in a safe, sanitary and efficient manner.

Regulatory Considerations

When planning facilities for food establishments, you must know about and comply with national, state, and local standards and codes related to:

◆ Health
◆ Safety
◆ Building
◆ Fire
◆ Zoning
◆ Environmental code standards.

KEY TERMS

Equipment - the appliances: stoves, ovens, etc.; and storage containers, such as refrigerated units used in food establishments.

The **equipment** used in food establishments should meet the American National Standards Institute (ANSI) standards and bear the stamp of approval of recognized third-party certification organizations. Some examples of these organizations include:

◆ NSF International
◆ Underwriters Laboratories, Inc. (UL)
◆ American Gas Association (AGA).

Buyers who purchase equipment approved by these organizations are assured quality materials are used in the construction. Also, the items are designed and constructed to meet accepted food sanitation criteria.

UL and NSF seals

Work Center Planning

The production areas in a food establishment are commonly organized into work centers. These are areas where a group of closely related tasks are performed by an individual or individuals. The number of work centers required in a food establishment depends on the number of functions to be performed and the volume of material to be handled.

An employee should be able to complete the related tasks at the work center without moving away from it. The work center should also be large enough to do the job yet small enough to reduce travel and conserve time and effort. A properly designed work center will provide adequate facilities and space for:

◆ Efficient production
◆ Fast handling and service
◆ A pleasant environment
◆ Effective cleanup.

Equipment Selection

It is extremely important to select the right piece of equipment for the job. Compare different pieces of equipment for a particular job.

Features to consider when selecting equipment

◆ Design   ◆ Size

◆ Construction   ◆ Cost

◆ Durability   ◆ Safety

◆ Ability to clean easily   ◆ Overall ability to do the job.

Purchase equipment that will improve the quality of food, reduce labor and material costs, improve sanitation, and contribute to the bottom line of the establishment.

KEY TERMS

Easy to clean - materials and design which facilitate cleaning of equipment.

Size and Design

Design is an important feature of food equipment, since the equipment used in a food establishment is subject to constant use and abuse. Equipment and utensils must be designed to function properly when used for their intended purposes.

Types of equipment

Floor-mounted Equipment	Counter-mounted Equipment

Floor-mounted equipment must be:	Counter-mounted equipment must be:
◆ Elevated on 6-inch (15 cm) legs, or;	◆ Elevated on 4-inch (10 cm) legs, or;
◆ Sealed to the floor, or;	◆ Sealed to the counter.
◆ Mounted on casters to make it easily movable.	
Clearance space and mobility make it easier to clean the floor under and behind the equipment. Equipment sealed to the floor will prevent the accumulation of debris and the harborage of pests.	This provides clearance between the counter top and the bottom of the equipment and makes it easier to clean under and around the equipment. Waste will not collect under equipment that is sealed to the counter.

Construction Materials

The *Food Code* and construction standards, such as those from ANSI, require food equipment and utensils to:

◆ Be smooth
◆ Be seamless
◆ Be easily cleanable
◆ Be easy to take apart
◆ Be easy to reassemble
◆ Have rounded corners and edges.

Materials used in the construction of utensils and food-contact surfaces of equipment must be nontoxic and not impart colors, odors, or tastes to foods. Under normal use, these materials must also be safe, durable, corrosion-resistant, and resistant to chipping, pitting, and deterioration.

Types of materials

Metal

- Popular materials in food establishments
- Commonly used in conjunction with small appliances
- Noncorrosive metals formed by the alloys of iron, nickel, and chromium may also be used in the construction of food equipment.

Advantages:

- Chromium over steel gives an easily cleanable, high-luster finish.

Disadvantages:

- Lead, brass, copper, cadmium, and galvanized metal can cause a chemical poisoning when they come into contact with high-acid foods (foods that have a low pH). Therefore, these materials must not be used as food-contact surfaces for equipment, utensils, and containers.

Stainless Steel

- Stainless steel is one of the most popular materials in food establishments. It is commonly the material of choice for food containers, counter tops, sinks, dish tables, dishwashers and ventilation hood systems.
- Finish No. 4 (on a scale of 1-8) is most commonly preferred for food production areas whereas higher finishes are usually preferred in display and serving areas.

Advantages:

- Has a durable, shiny surface that easily shows soil and is easy to clean and maintain
- Resists high temperatures, rust, and stain formation.

Disadvantages:

- Higher cost
- Extensive polishing required
- Polishing requires labor, material, and energy.

(cont.)

Abrasive cleaners and scouring pads can scratch the surface of the metal and cause harborage areas for disease-causing microorganisms.

Plastic

You must be sure to buy food equipment that is made of only food-grade plastics. Select the one that works best for you based on intended use and durability.

The harder, more durable plastics are easier to clean and sanitize. Some examples of plastics used in food establishments are:

- Acrylics (used to make covers for food containers)
- Fiberglass (used in boxes, bus trays, and trays)
- Polyethylene (used in storage containers and bowls).

Courtesy of Ecolab Food Safety Specialties

Advantages:

- Durable
- Inexpensive
- can be molded into different combinations

Wood

- The *Food Code* permits limited use of wood materials including hard maple or an equally hard, close-grained wood for cutting boards, cutting blocks, and baker's tables. Wood is also approved for paddles used in pizza operations.

Advantages:

- Light in weight
- Economical.

Disadvantages:

- Porous to bacteria and moisture
- Absorbs food odors and stains
- Wears easily under normal use
- Requires frequent maintenance and replacement.

The consensus seems to be the disadvantages of using wood for food-contact surfaces outweigh the advantages. This is primarily because of the high degree of maintenance required and problems with keeping the equipment clean and sanitary.

Types of Equipment

NSF International and Underwriters Laboratories, Inc., provide independent evaluations of equipment and materials. The NSF and UL emblems on a piece of equipment verify the equipment has been tested and meets the requirements prescribed in their standards.

Cooking Equipment

The most important criteria to use when selecting cooking equipment are the types and quantities of food prepared, ease of cleaning, durability and energy conservation. The frame, door, exterior, and interior materials of cooking equipment should contribute to the durability and cleanability of the equipment. The type and thickness of insulation materials contribute to the energy efficiency of this equipment.

Ovens and Fryers

Ovens and fryers are among the most important pieces of equipment in a food establishment. They are used to cook a variety of foods to different temperatures. Heat is distributed by radiation, conduction, or convection, depending on the type equipment being used.

Cooking equipment should be:

- The right size and capacity
- Easy to clean
- Durable
- Energy efficient.

Range

Heat distribution: conduction

- Commonly used in small operations
- Cooking surface on top of oven.

Deck

Heat distribution: conduction

- Multiple ovens stacked on top of one another
- Each oven contains separate heating elements.

A good oven should:

- Rise to 450°F (232°C) within 20 minutes
- Have proper heat circulation
- Be able to cool quickly when a drop in temperature is required.
- Be well insulated to prevent heat loss.

Convection

Heat distribution: convection

- High-speed fan circulates heated air around food to reduce cooking time
- Multiple racks allow for more cooking in a smaller space.

Fryer

Heat distribution: conduction

- Commonly used in food production for standard menus.

(cont.)

Ovens (continued)

Microwave

Heat distribution: radiation

◆ Used for thawing, heating, and reheating foods
◆ Cooks small quantities of food quickly.

Other types of ovens may be used in food establishments. These include rotary, infrared, conveyor, and roll-in units. Each of these pieces of equipment has unique features and has been designed for special applications. You should contact your equipment supplier to determine if this type equipment is best suited to carry out the functions required in your operation.

Refrigeration and Low Temperature Storage Equipment

Refrigerators and freezers are used to keep perishable foods fresh and preserve the safety and wholesomeness of PHF (TCS). Many foods deteriorate rapidly at room temperature. Food establishments can reduce spoilage, waste, and shrinkage by keeping foods at lower temperatures until they are used.

Proper cooling requires removing heat from food quickly enough to prevent the growth of spoilage and disease-causing microbes. Improper cooling of PHF (TCS) products has been consistently identified as a leading contributor to foodborne illness. Bacteria grow best at temperatures between 70° (21°C) and 120°F (49°C). Therefore, the *Food Code* states cooked PHF (TCS) must be cooled from 135°F (57°) to 70° (21°C) within two hours and from 135°F (57°C) to 41F (5°F) or below within six hours.

All types of refrigerators and freezers have a maximum capacity for cooling foods. Too often, employees put large amounts of hot food in a unit and rely on it to "cool" the food. On the contrary, the addition of large amounts of hot food causes the inside temperature of the unit to rise above acceptable storage temperatures. This, in turn, causes food in the unit to be stored in the temperature danger zone. Don't forget the primary purpose of a refrigerator is to take foods that are already at or near the proper cold-holding temperature and keep them out of the temperature danger zone.

> **The primary purpose of a refrigerator is to take foods that are already at or near the proper cold-holding temperature and keep them out of the temperature danger zone.**

IMPORTANT!

According to Food Code:

Food Code
U.S. Public Health Service
FDA
2009

The *Food Code* states cooked PHF (TCS) must be cooled from 135°F (57°) to 70° F(21°C) within two hours and from 135°F (57°C) to 41°F (5°F) or below within six hours.

The efficient operation of a refrigeration unit depends on several factors including:

◆ Design
◆ Construction
◆ Capacity of the equipment.

Do not exceed the cooling capacity of your refrigeration equipment and avoid stocking food above the maximum load line on display equipment.

Refrigerated prep-table

Proper construction of cold storage units includes sturdy construction of doors, hardware, and fixtures. Doors may be full-length or half-length and shelving should be adjustable. Door gaskets should be durable and easy to clean. They must be replaced when worn.

The size of refrigerator or freezer needed depends on the size of the work area and the type and amount of food to be stored in the unit. Refrigeration and low temperature storage equipment must be adequately sized and properly installed to assure reliable and efficient operations.

> **Proper air circulation both inside and outside the refrigerator makes it more efficient.**

Maintenance of the refrigeration equipment in the various work areas is an important responsibility. These units must be cleaned on a regular basis to maintain good sanitary conditions and eliminate odors. The inside walls, floor, shelves, and other accessories of a walk-in refrigeration unit must be cleaned regularly to remove spills and debris. Don't forget to clean the fan grates and condenser as part of your routine cleaning.

✔ DO	⊘ DON'T
✔ Store foods on slatted shelves to permit good circulation of air	⊘ Line shelves with sheet pans, foil, plastic, or cardboard - this decrease the flow of cool air
✔ Refrigerate foods that are already at or near the proper cold-holding temperature	⊘ Put large amounts of hot foods directly into the refrigerator or freezer to cool
✔ Cool foods in shallow containers	⊘ Put large amounts of solid or semi-solid hot food directly into the cooler
✔ Keep units clean, including fan grates and condensers	⊘ Forget to clean the unit on a regular basis
✔ Store raw foods under cooked and ready-to-eat foods to avoid cross contamination.	⊘ Place raw foods over ready-to-eat.

Reach-In Refrigeration

Reach-in

Courtesy of Hobart Corp.

- Models range in capacity
- Can have multiple doors
- May have external thermometer attached to a sensing device inside the warmest part of the refrigeration unit.

Walk-in

- Coolers are used to store large quantities of perishable foods between 32°F (0°C) and 41°F (5°C)
- Used to thaw products
- Can be combined with reach-in dairy/deli display cases that are loaded from inside the walk-in
- Door openings may have 4-inch wide plastic strips called strip curtains which reduce cool air loss when the door is open
- Freezers are used to keep large quantities of frozen foods solidly frozen
- May have external thermometer attached to a sensing device placed at the warmest location in the refrigeration unit.

Display

- Keep cold foods out of the temperature danger zone
- Do not exceed the cooling capacity of your refrigeration equipment
- Avoid covering vents and return air openings
- Properly control defrost cycles
- May be open air or have closing doors.

Cook-chill and Rapid-chill Systems

Cook-chill is a system in which food is:

◆ Cooked using conventional cooking methods
◆ Rapidly chilled using a chiller
◆ Stored for a limited time
◆ Reheated before service to the customer.

Rapid-chill equipment

The food is typically chilled to 37°F (3°C) in 90 minutes or less and is stored at temperatures between 33°F (1°C) and 38°F (3°C).

The advantages of this system include reduction of peaks and valleys in production and readily available foods. The disadvantages of this system include the high cost of the chiller and the storage space requirements.

Rapid-chill systems are designed to cool hot foods very quickly. This type of equipment can typically get a few hundred pounds of hot food through the temperature danger zone in 2 hours or less. Although this equipment is somewhat expensive, it can be a very good investment for some food establishments that work with large masses of food or with foods that are challenging to cool quickly.

Be prepared in case of a power failure:

◆ Keep refrigerator doors closed
◆ Monitor product temperatures
◆ Discard products that are in the temperature danger zone for more than 4 hours.

Hot-Holding Equipment

PHF (TCS) products that have been cooked and are to be eaten hot must be held at 135°F (57°C) or above. An important fact to remember is hot-holding equipment will not raise the temperature of foods very much. In order for hot-holding equipment to work properly, the food must be at 135°F (57°C) or above when it is put into or onto this equipment.

Hot-holding equipment uses steam, heating elements, and/or light bulbs to keep foods hot. This equipment should be checked regularly to make sure it is working properly. Food temperatures must be monitored to assure they are being maintained at 135°F (57°C) or above. Hot-holding units must be cleaned regularly, and food should be rotated using a FIFO procedure.

Hot-holding equipment is not designed to raise the temperature of foods. Food must be at 135°F (57°C) or above when it is put into or onto this equipment.

Hot-holding equipment

Other Types of Food Equipment

Slicers

The basic design of food slicers includes a circular knife blade and carriage that passes under the blade. Foods to be sliced are placed on the carriage and fed either manually or automatically. Slicers can be dangerous if not used properly. The manufacturer's instructions should always be followed, and employees should receive training on safe operation and proper cleaning of this equipment.

Traditional design

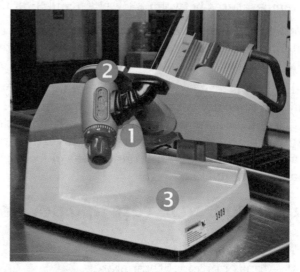

New design to facilitate sanitation

Slicers have been redesigned to be easier to clean and sanitize.
(See photo above)

1 Moved red index knob out of food splash zone

2 Enlarged space between blade area and motor housing for easier accessibility and cleaning.

3 One-piece base eliminates seams and potential collection areas.

Meat grinder

Courtesy of Hobart Corp.

Mixers, grinders, and choppers

Mixers, grinders, and choppers are examples of equipment commonly found in a food establishment. Some of these pieces of equipment are available as counter and floor models.

Mixers can also be used to shred and grind when different accessories and attachments are used. Floor model mixers have three standard attachments: (1) a paddle beater for mixing or blending foods and ingredients; (2) a whip to incorporate air into products; and (3) a dough hook used to mix and knead dough.

Grinders and choppers work well with a variety of fresh foods and ingredients including meats and vegetables.

Ice machines

Ice is an important item in food establishments. It is used to chill beverages and preserve the freshness of fish and some produce items during display. Ice must be made from potable water, and ice machines must protect the ice during production and storage.

The parts of an ice machine that come into contact with the ice must be smooth, durable, easily cleanable and constructed of nontoxic materials. These food-contact surfaces must be cleaned and sanitized regularly to prevent the growth of mold and other microorganisms. The drain line from the ice machine must be equipped with an air gap to protect the ice from contamination due to backflow. You will learn more about backflow and air gaps in Chapter 7.

Ice machine

Scoops, shovels, carts, and other equipment used to dispense or transport ice must meet the design and construction criteria for food-contact surfaces. When not in use, this equipment should be stored in a manner that will protect it from contamination.

Employees must never dispense ice by passing a glass or cup through the ice. This can cause glass from the container to break off and become a physical hazard in the ice. Food and beverage containers must not be stored in ice that will be used for drinking purposes. This prevents contamination of the ice your customers may consume. When scoops are stored inside an ice machine, you must ensure the handle of the scoop extends above the ice. This will permit employees to use the handle of the scoop without their hands touching the ice or the food-contact surface of the scoop.

> The drain line of an ice machine must be equipped with an air gap to prevent contamination due to backflow.

> Never allow employees to use a glass or cup to dispense ice. Always use an ice scoop.

> Do not store food and beverage containers in ice served to customers.

Single-Service and Single-Use Articles

Single-use articles include tableware, carryout utensils, and other items such as bags, containers, stirrers, straws, and wrappers designed and constructed to be used only one time by only one person. After one use the article is discarded.

Single-use articles include items such as wax paper, butcher paper, deli paper, plastic wrap, and certain types of food containers that are designed to be used once and discarded.

A food establishment should provide single-use and single-service articles for food handlers if it does not have proper facilities for cleaning and sanitizing multi-use kitchenware and tableware. These establishments must also provide single-service articles for use by consumers.

Materials used to make single-service and single-use articles must not permit the transfer of harmful substances or pass on colors, odors, or tastes to food. These materials must be safe and clean when used in food establishments.

Warewashing Equipment

Equipment and utensil dishwashing should be performed in a room or area separate from food production areas.

Warewashing is the process used to clean and sanitize the equipment, utensils, dishes, glasses, etc., that are used during the preparation, handling, and consumption of foods. Proper warewashing is one of the most important jobs in a food establishment. As you learned in Chapter 3, contaminated equipment and utensils have been identified as sources of contamination and cross contamination that can cause foodborne illness.

Warewashing areas must be well lighted and well ventilated. Noise absorbing materials may be installed on walls and the ceiling to lower noise levels in warewashing areas.

Some of the items most frequently washed and sanitized in a food establishment are:

◆ **Utensils** such as knives, forks, spoons, and tongs

◆ **Kitchenware** such as pots, pans, cutting boards, slicers, grinders, and mixers

◆ **Tableware** such as dishes, glasses, and eating utensils.

The purpose of warewashing is to clean and sanitize equipment and utensils. It consists of two phases:

1. A cleaning phase where visible soil is removed from the surface of the item through washing and rinsing.

2. A sanitizing phase where the number of disease-causing microorganisms on a cleaned surface is reduced to safe levels.

Cleaning and sanitizing operations can be performed either manually or mechanically.

Manual Warewashing

A manual warewashing area must provide adequate space to store soiled equipment and utensils. Items to be cleaned must be pre-flushed or pre-scraped and, if necessary, pre-soaked to remove food particles and soil. A hose and nozzle or other device must be provided to pre-flush and pre-scrape food soil into a garbage container or disposal. This equipment must be located at the soiled end of the warewashing operation to avoid contaminating cleaned and sanitized equipment and utensils.

Manual warewashing can also be performed using a bucket and brush, wall-mounted hose units, and spray units. These processes will be explained in more detail in Chapter 6 of this book.

Three-compartment sink

The compartments of the three- or four-compartment sink must be large enough to accommodate the largest pieces of equipment and utensils used in the food establishment. Supply each compartment with hot and cold potable running water. Provide drainboards or easily movable dish tables of adequate size for proper handling of soiled utensils prior to washing, and for air-drying cleaned and sanitized items.

Mechanical Warewashing

Warewashing machines are used to clean and sanitize equipment and utensils that do not have electrical parts and that will fit into the machine. When using a single-tank, stationary-rack dishwashing machine, the dishes are placed on racks and washed one rack at a time with jets of water within a single tank. They are operated by opening a door, inserting a rack of dishes, closing the door, and starting the machine.

A dishwashing machine must automatically dispense detergents and sanitizers. There must be a visual means to verify detergents and sanitizers are delivered or a visual or audible alarm to signal if the detergent and sanitizers are not delivered to the respective washing and sanitizing cycles. The machine must be large enough to accommodate the size and volume of equipment and utensils to be cleaned and sanitized.

Low temperature dishwashers are similar in design to the single-tank, stationary-rack dishwashing machine. However, they use chemicals to sanitize equipment and utensils. This allows lower water temperatures, which conserves energy.

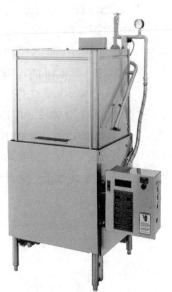

Sngle-tank dishwashing machine

(Courtesy of Hobart Corp.)

When purchasing a dishwashing machine, you should also consider the cost of operation and maintenance. An adequate supply of very hot water is required for the final rinse in a high temperature dishwashing machine. The high temperatures required for effective sanitization make it expensive and unsafe to maintain general purpose hot water at these temperatures. Therefore, a separate booster heater is needed to raise the temperature of the sanitizing rinse water to the proper temperature. The booster heater must be properly sized, installed, and operated to deliver water at the volume, flow pressure, and temperature required for the operation.

> **The most common types of mechanical dishwashers used in retail food establishments are:**
>
> ◆ Single-tank
> ◆ Stationary-rack
> ◆ Low temperature.

Installation

Proper installation is required to assure equipment functions properly. The best design and construction will be worthless if electrical, gas, water, or drain connections are inadequate or improperly installed. The dealer who sells you equipment may or may not be responsible for its installation. Arrangements for installation will usually be specified in your purchase agreement. Following installation, employees must be trained to operate equipment correctly and safely.

Maintenance and Replacement

The cost of care and upkeep on a piece of equipment may determine whether or not its purchase and use are justified. Successful maintenance of equipment requires definite plans to prolong its life and maintain its usefulness.

Maintain your equipment

- ◆ Keep the equipment clean
- ◆ Follow the manufacturer's printed directions for care and operation
- ◆ Post the instruction card for a piece of equipment near it
- ◆ Stress careful operation and maintenance schedules
- ◆ Make needed repairs promptly.
- ◆ Some valuable suggestions for the care of equipment are:
  - ◆ Assign the care of a machine to a responsible person
  - ◆ Check the cleanliness of machines daily
  - ◆ Have repairs performed promptly and by a properly trained person.

Lubricants used on food equipment may directly or indirectly end up in food. Therefore, all lubricants used on food-contact surfaces, bearings and gears, and other components of equipment that are located so the lubricants may leak, drip or be forced into food or onto food-contact surfaces must be approved as food additives or generally recognized as safe. These lubricants should be used in the smallest amount needed to get the job done.

Lighting

Proper lighting in production and dishwashing areas:

- ◆ Increases productivity
- ◆ Improves workmanship
- ◆ Reduces eye fatigue and employee irritability
- ◆ Decreases accidents and waste due to employee error.

Glass and food do not mix.

(Courtesy of Shat-R-Shield, Inc.)

Food production and warewashing areas should be furnished with the proper amount of lighting and soft colors that reduce glare. Proper lighting also shows soil and when a surface has been cleaned. Locate lights to eliminate shadows on work surfaces or glare and excessive brightness in the field of vision.

The amount of light required in a work area depends on the kind of work performed there. The following light intensity levels are recommended in the *Food Code.*

Recommended light intensity levels

10 foot-candles at a distance of 30 inches above the floor

◆ Walk-in refrigeration units
◆ Dry food storage areas
◆ Other areas and rooms during periods of cleaning.

20 foot-candles at a distance of 30 inches above the floor

◆ Areas where fresh produce and packaged foods are offered for sale and consumption
◆ Areas used for hand washing
◆ Areas used for warewashing
◆ Areas used for equipment and utensil storage
◆ Toilet rooms.

50 foot-candles at the work surface

◆ Where an employee is working with unpackaged, PHF (TCS) products
◆ Where an employee is working with food, utensils, and equipment such as knives, slicers, grinders, or saws
◆ Where employee safety is an overriding concern.

The food regulations in some state and local jurisdictions may require higher illumination levels in food establishments, especially during periods of cleaning. Consult the food regulation in your jurisdiction to determine what lighting levels are required for your establishment.

Light bulbs must be shielded, coated or otherwise shatter-resistant when used in areas where there is exposed food; clean equipment, utensils, and linens; or unwrapped single-service and single-use utensils. This is to prevent glass fragments from getting into food and onto food-contact surfaces. Shielded or shatter-resistant bulbs are not required in locations where stored packaged foods will not be affected by broken glass falling on them because the packages can be cleaned of debris from broken bulbs before being opened.

Light bulbs over self-service buffets and salad bars must also be shielded or shatter-resistant. In addition, this lighting must not mislead customers about the quality of the food items on display.

Plastic-coated shatterproof bulbs
(Courtesy of Shat-R-Shield, Inc.)

Heating, Ventilation, and Air Conditioning (HVAC)

Air conditioning in food establishments means more than simply "cooling the air." It includes heating, humidity control, circulation, filtering, and cooling of the air. HVAC systems must filter, warm, humidify, and circulate the air in the winter and maintain comfortable air temperature in the summer.

Ventilation in food production and warewashing areas is typically provided by mechanical exhaust hood systems. These systems keep rooms free of excessive heat, steam, condensation, vapors, obnoxious odors, smoke, and fumes. The standard ventilation system used in food establishments consists of a hood, fan, and intake and exhaust air ducts and vents. Ventilation hood systems must be designed and constructed to prevent grease or condensation from dripping onto food, equipment, utensils, linens, or single-service and single-use articles. The capacity of the ventilation system should be based on the quantity of vapor and hot air to be removed.

> **Ventilation systems keep rooms free of excessive heat, steam, condensation, vapors, obnoxious odors, smoke, and fumes.**

Hoods are usually constructed of stainless steel or a comparable material that provides a durable, smooth, and easily cleanable surface. The hood should be equipped with filters or other grease extracting equipment to prevent drippage onto food.

Filters and other grease removal equipment must be:

◆ Easily removed for cleaning and replacement
◆ Designed to be cleaned in place.

Fires at food establishments have been caused by the combustion of grease that accumulates in filters and ducts. Cleaning and replacement of filters must be part of your routine maintenance program. Intake and exhaust air ducts must be cleaned so they do not become a source of contamination of dust, dirt, and other materials. If vented to the outside, ventilation systems must not create a public health nuisance or unlawful discharge.

What Do You Think?

Defective lighting poses serious problem at local deli . . .

A fluorescent light bulb burned out in the food preparation area at Mike's deli. The fixture was located immediately above a work counter used by employees to transfer cold meat salads from bulk containers into shallow containers for display.

A maintenance man came to replace the bulb that was covered by a plastic shield. Just as he pulled the bulb out, the shield split and small pieces of glass fell into the food containers.

What problems have just occurred at the deli counter? What would you do in this

Summary

Back to the Story... a fluorescent bulb, even though shielded, split open when the maintenance employee removed the object. Broken glass fell into open containers of food. The food in the containers was contaminated and had to be discarded. To prevent similar situations in the future, maintenance personnel must be instructed not to replace burned out light bulbs while food-handling activities are being conducted on the counter below. If it was necessary to replace the bulb immediately, the deli employees should have removed the food and containers from the counter area.

Good planning is essential for maximum productivity and efficiency. The total allotted space needed and the arrangement of equipment in the space are the overall features to consider when planning the layout of a food establishment. The design of the facility should promote the smooth flow of work, provide adequate facilities and space for performing work efficiently, and contribute to high sanitation standards.

The basic needs of a food establishment will determine the type and size of equipment purchased. When purchasing equipment, look for items that:

◆ Will serve the current and future needs of the operation
◆ Can be properly cleaned and sanitized
◆ Do not require extraordinary maintenance and repair.

The *Food Code* and construction standards from the American Standards Institute (ANSI) require food equipment and utensils to be:

◆ Smooth
◆ Easily cleanable
◆ Easy to take apart
◆ Easy to reassemble
◆ Equipped with rounded corners and edges.

Materials used as food-contact surfaces must be safe, durable, corrosion-resistant, nonabsorbent, and resistant to chipping, pitting, and deterioration. These materials must not allow the transfer of harmful substances, colors, odors, or tastes to food.

Always install equipment in accordance with local building, plumbing, electrical, health, and fire safety codes. The best possible design and construction of equipment is worthless if electrical, gas, water, or drain connections are inadequate or the equipment is poorly installed.

Quiz 5 (Multiple Choice)

Please choose the BEST answer to the questions.

1. Shelves that are mounted to the floor must have at least _____ inches of clearance underneath to permit easy cleaning and discourage pest harborage.

    a. 12.
    b. 10.
    c. 8.
    d. 6.

2. Counter-mounted equipment that is not easily movable or sealed to the counter must be at least _____ inches above the counter to permit easy cleaning under and behind the equipment.

    a. 2.
    b. 4.
    c. 6.
    d. 8.

3. Which of the following metals is approved for food-contact surfaces?

    a. Stainless steel.
    b. Lead.
    c. Copper.
    d. Galvanized metal.

4. From a **sanitation** perspective, what is the most important reason for having adequate lighting in a food production area?

    a. To provide a comfortable work environment for employees.
    b. To show when a surface is soiled and when it has been properly cleaned.
    c. To decrease accidents and waste due to employee error.
    d. To reduce glare that causes eye fatigue.

5. Which of the following statements about ventilation is **false**?

    a. Ventilation is typically provided by means of a mechanical exhaust hood system.
    b. Ventilation hood systems must be designed and constructed to prevent grease or condensation from dripping onto food and food-contact surfaces.
    c. The hood should be equipped with filters or other grease-catching devices to prevent drippage into food.
    d. Intake and exhaust air ducts do not need cleaning if filters or other grease catching devices are provided to collect grease and other condensation.

6. Which of the following statements about ice and ice machines is **false**?

   a. Ice is food and must be made from potable water.

   b. It is permissible to store bottles of food and beverages in ice that will be served to customers.

   c. Scoops used to dispense ice must be stored in a way that will prevent employees' hands from touching the ice.

   d. Harmful microbes can survive on ice.

7. Which of the following statements about food equipment is **false**?

   a. Equipment should be properly sized so it can be installed in the space available and be easily cleaned.

   b. Construction requirements for equipment vary according to whether the surface or area is a "food-contact surface" or a splash zone (non-food-contact area).

   c. Wood is a preferred material for food-contact surfaces because it resists moisture and bacterial growth.

   d. Counter-mounted equipment that is not easily movable must be elevated on 4-inch legs or sealed to the counter.

8. Which of the following statements about refrigeration equipment is **false**?

   a. The size of refrigerator or freezer needed depends on whether a walk-in unit is available, what is going to be stored, and how much.

   b. All types of refrigerators have a maximum capacity for cooling foods.

   c. The primary purpose of a refrigerator is to take foods that are already at or near the proper cold-holding temperature and keep them out of the temperature danger zone.

   d. The cleanliness and maintenance of refrigeration equipment is not important since this equipment is only used to store packaged foods.

9. What type of equipment does a food establishment use when conducting manual warewashing operations?

   a. A 3- or 4-compartment sink equipped with hot and cold potable, running water and drainboards or easily moveable dish tables.

   b. A 2-compartment sink equipped with hot and cold potable, running water and drainboards or easily moveable dish tables.

   c. A 2-compartment sink equipped with suitable facilities for pre-flushing and pre-scraping items before they are washed.

   d. A low temperature, chemical sanitizing dishwashing machine that is large enough to accommodate the largest pieces of equipment and utensils used in the operation.

10. Which of the following statements about mechanical warewashing equipment is **false**?

   a. Dishwashing machines can be used to clean and sanitize equipment and utensils that do not have electrical parts.

   b. Dishwashing machines must automatically dispense detergents and sanitizers.

   c. A dishwashing machine must provide a visual means to verify detergents and sanitizers are delivered or a visual or audible alarm to signal if the detergent and sanitizers are not delivered to the respective washing and sanitizing cycles.

   d. Low temperature dishwashing machines use hot water at 140°F (60°C) to sanitize equipment and utensils.

Answers to the multiple-choice questions are provided in **Appendix A**.

References/Suggested Readings

Baraban, Regina S. and Joseph F. Durocher. 2010. *Successful Restaurant Design*. John Wiley and Sons, New York, NY.

Birchfield, John C. and John Birchfield, Jr. 2008. *Design and Layout of Foodservice Facilities*. John Wiley and Sons, Hoboken, NJ.

Food and Drug Administration and Conference for Food Protection. 1997. *Food Establishment Plan Review Guide*, Washington, D.C.

Food and Drug Administration. 2009. *2009 Food Code* - Chapter 4. U.S. Public Health Service, Washington, D.C.

Food and Nutrition Service. 1999. *A Guide for Purchasing Food Service Equipment*. U.S. Department of Agriculture, Washington, D.C.

Fullen, Sharon L. 2003. The Food Service Professionals Guide to Restaurant Design: Designing, Constructing & Renovating a Food Service Establishment. Atlantic Publishing Group, Ocala, FL.

Katsigris, Costas and Chris Thomas. 2006. Design and Equipment for Restaurants and Foodservice: A Management View. John Wiley and Sons. New York, NY.

Longrèe, K., and G. Armbruster. 1996. *Quantity Food Sanitation*. Wiley Interscience, New York, NY.

Suggested Web Sites

AK Steel Stainless Steel
   www.aksteel.com/markets_products/stainless.aspx
NSF International
   www.nsf.org
Underwriters Laboratories, Inc.
   www.ul.com
Hobart Corp.
   www.hobartcorp.com
Hussmann Co.
   www.hussmann.com
Katchall Industries International
   www.KatchAll.com
Shat-R-Shield, Inc.
   www.shat-r-shield.com
Specialty Steel Industry of North America
   www.ssina.com
The Food Marketing Institute
   www.fmi.org

The National Restaurant Association

www.restaurant.org

Gateway to Government Food Safety Information

www.foodsafety.gov

San Jamar

www.sanjamar.com

Chapter 6

Cleaning and Sanitizing Operations

Key Terms

- Chlorine
- Cleaning agent
- Cleaning
- Clean-in-place (CIP)
- In-place sanitizers
- Iodophors
- Potable water
- Quaternary ammonium (quats)
- Sanitizers
- Sanitizing

Cleaning and sanitizing are important activities in all food establishments.
Proper cleaning and sanitizing enhance the safety and quality of food, increase
the life expectancy of equipment and facilities, and improve overall sanitary
conditions.

Learning Points:
◆ Recognize the difference between cleaning and sanitizing.
◆ Recognize the relationship between cleaning and sanitizing.
◆ Identify the different processes that can be used to clean and sanitize equipment
   and utensils in a food establishment.
◆ Identify the primary steps involved in manually and mechanically cleaning and
   sanitizing equipment and utensils.
◆ Describe the factors that affect cleaning efficiency.
◆ Identify the procedures used to clean environmental areas in a food
   establishment.

Principles of Cleaning and Sanitizing

Cleaning and sanitizing are two distinct processes used for very different purposes. Cleaning is the physical removal of soil from surfaces of equipment and utensils. Sanitizing or sanitization is the treatment of a clean surface to reduce the number of disease-causing microorganisms to safe levels. Most of the soil is food wastes and residues. The equipment and supplies used for cleaning are different from those used for sanitizing. Single-use items do not need to be cleaned because they must be discarded after use.

Water is the primary component of the cleaning materials used in food establishments. The water supplied to a food establishment must be **potable** or safe to drink. While the water supply must be potable, it may contain impurities that cause hardness, taste, and odors. Therefore, your cleaning agents must work well in the type of water that is supplied to your establishment. See Appendix H for more information about factors that effect cleaning and sanitizing.

KEY TERM

Potable water- drinking water that meets national drinking water regulations and is safe to drink.

Cleaning and Sanitizing are Two Different Things...

Cleaning

Cleaning is the physical removal of soil from surfaces of equipment and utensils.

Sanitizing

Sanitizing or sanitization is the treatment of a clean surface to reduce the number of disease-causing microorganisms to safe levels.

You must clean equipment and utensils before you can sanitize them.

The cleaning and sanitizing processs in food establishments consists of 5 steps:

Pre-scrape → Wash → Rinse → Sanitize → Air-Dry

Cleaning Principles

Effective cleaning consists of five separate events:	
Pre-scrape/Pre-flush	◆ Scrape and flush food particles from the surfaces of equipment and utensils before the items are placed in a cleaning solution.
Apply cleaning agent	◆ A detergent or other type of cleaner is brought into contact with the soil.
Loosen soil	◆ The soil is loosened from the surface being cleaned.
Disperse the soil in water	◆ The loosened soil is dispersed in the wash water.
Rinse	◆ The dispersed soil is rinsed away along with the detergent to prevent it from being redeposited onto the clean surface.

Pre-scrape / Pre-flush to Remove Food Particles

Removal of food particles is critical for effectve cleaning. Scrape and flush food particles from the surfaces of equipment and utensils before the items are placed in a cleaning solution. Use warm water when pre-scraping and pre-flushing equipment and utensils. Avoid using very hot water or steam because they tend to "bake" food particles onto the surface of equipment and utensils and that makes cleaning more difficult.

Pre-scraping makes washing easier.

Application of Cleaning Agents

A **cleaning agent** is a chemical compound formulated to remove soil. It is important to choose a cleaning agent that is right for the particular job you have. There are many methods of applying cleaning agents to surfaces of equipment.

KEY TERM

Cleaning agent- a chemical compound formulated to remove soil and dirt.

Soaking

Small equipment, equipment parts, and utensils may be immersed in cleaning solutions in a sink. By soaking equipment or utensils for a few minutes before scrubbing, you will increase the effectiveness of manual and mechanical warewashing.

Spray Methods

Cleaning solutions can be sprayed on equipment surfaces by using either fixed or portable spray units that use hot water or steam. These methods are used extensively in meat departments of food establishments.

Clean-in-place systems

The clean-in-place method is an automated cleaning system generally used when cleaning pipeline systems such as those found in soft serve ice cream dispensing equipment. The strength and velocity of the cleaning solution moving through the pipes are chiefly responsible for removing soil in clean-in-place (CIP) operations.

Spray methods are used extensively in meat departments.

Abrasive cleaning

Abrasive cleaners, in the form of powders and pastes, are used to remove soil firmly attached to a surface. Always rinse these cleaners completely and avoid scratching the surface of equipment and utensils. Abrasive type cleaners are not recommended for use on stainless steel surfaces. Never use metal or abrasive scouring pads on food-contact surfaces because small metal pieces from the pads may promote corrosion or may be picked up in food to become a physical hazard.

Detergents and Cleaners to Be Used

The origin of the word detergent is from the Latin, *detergeo*, meaning "to wipe away." Water acts as a detergent when soils are readily soluble. However, we can improve the cleansing action of water by adding soap, alkaline detergents, acid detergents, degreasers, abrasive cleaners, detergent sanitizers, or other cleaning agents to it. See Appendix H for a list of different types of detergents and the advantages and disadvantages they provide.

Factors that affect cleaning efficiency

1. **Type of and amounts of soil to be removed–** determines the cleaning agents and process used to remove soil.

2. **Water quality–** Water must be potable and compatible with the water supply used at the food establishment.

3. **Detergent or cleaner to be used–** Must be compatible with the type of soil and water supply of the food establishment.

4. **Concentration of cleanser–** Using the recommended amount of detergent improves cleaning power.

5. **Amount of time detergent/cleaner remains in contact with the surface–** Reduces scrubbing necessary to remove soil.

6. **Water temperature–** Increased water temperature helps decrease the strength of the bonds that hold soil to the surface. Avoid water that is too hot and will bake soil onto surface of equipment and utensils.

7. **Velocity or force–** Removes soil and film from food-contact surfaces.

Rinsing

Immediately after cleaning, thoroughly rinse all equipment surfaces with hot, potable water to remove the cleaning solution. This very important rinse step is necessary because the soap or detergent used for cleaning can interfere with the effectiveness of the sanitizer.

> **If there is an extended interruption in the water supply to a food establishment, the facility should cease operations or sell only prepackaged or pre-prepared food using single-use utensils until water service is restored.**

Cleaning frequency

Under normal circumstances, food-contact surfaces and equipment used to prepare and serve PHF (TCS) must be cleaned throughout the day to prevent the growth of microorganisms on those surfaces. Some guidelines for cleaning food-contact surfaces include:

◆ Before each use with a different type of raw animal food such as beef, fish, lamb, pork, or poultry, except when the surface is in contact with a series of different raw animal foods each requiring a higher cooking temperature

◆ Each time there is a change from working with raw foods to working with ready-to-eat foods

◆ Between uses with raw fruits and vegetables and with PHF (TCS)

◆ Before using or storing a food temperature-measuring device

◆ At any time during the operation when contamination may have occurred.

There are some exceptions to the 4-hour cleaning rule. One exception is where equipment and utensils are used to prepare PHF (TCS) in a refrigerated room or area. The chart on the next page shows the cleaning frequency recommended in the *Food Code* for refrigerated areas.

According to Food Code:

◆ The *Food Code* requires food-contact surfaces of equipment and utensils used to prepare PHF (TCS) to be cleaned at least every 4 hours when used at room temperature.

◆ Food establishments must maintain records of the cleaning frequency based on the ambient temperature of the refrigerated room or area.

Room temperature and cleaning frequency

Room temperature	Cleaning frequency
◆ 41°F (5°C) or less	At least once every 24 hours
◆ 41°F (5°C) - 45°F (7°C)	At least once every 20 hours
◆ 45°F (7°C) - 50°F (10°C)	At least once every 16 hours
◆ 50°F (10°C) - 55°F (13°C)	At least once every 10 hours

In all cases, equipment and utensils that contact PHF (TCS) must be cleaned every 24 hours.

Iced tea dispensers, carbonated beverage dispenser nozzles, water dispensing units, ice makers and ice bins are examples of equipment that routinely come into contact with food that is not potentially hazardous. These types of equipment must be cleaned on a routine basis to prevent the development of slime, mold, or soil residues that may contribute to an accumulation of microorganisms.

Food Code
U.S. Public Health Service
FDA
2009

According to Food Code:

The *Food Code* recommends surfaces of utensils and equipment contacting food that is **not** potentially hazardous be cleaned:

◆ At any time when contamination may have occurred.

◆ At least every 24 hours for iced tea dispensers and customers' self-service utensils such as tongs, scoops, or ladles.

◆ Before restocking customers' self-service equipment and utensils such as condiment dispensers and bulk food display containers.

◆ Whenever possible, follow the manufacturer's guidelines for regular cleaning and sanitizing of the food-contact surfaces of equipment and utensils. If the manufacturer does not provide cleaning instructions, the person in charge should develop a cleaning regimen that will effectively remove soil, mold, and other contaminants from equipment and utensils.

Guidelines for cleaning food-contact surfaces that contact PHF (TCS)

Item and condition	When to be cleaned
◆ Storage containers for PHF (TCS) when properly maintained for hot- and cold-holding temperatures	◆ Clean when empty
◆ Containers in serving situations (salad bars, delis, cafeteria lines) which are periodically combined with same food refills at required temperatures	◆ At least once every 24 hours
◆ Temperature-measuring device maintained in contact with food (deli food or roast)	◆ At least once every 24 hours
◆ Refrigerated equipment used for storage of packaged or unpackaged food (reach-in refrigerator)	◆ At a frequency necessary to preclude accumulation of soil residue
◆ In-use utensils stored in water (containers with water at 135°F (57°C)	◆ At least once every 24 hours.

Sanitizing Principles

Heat and chemicals are the two types of sanitizer agents most commonly used in food establishments. Sanitizing does not provide the same level of microbial destruction as sterilization, because some bacterial spores and a few highly resistant vegetative cells generally survive the sanitizing process.

Heat Sanitizing

Heat sanitizing in manual warewashing operations requires cleaned equipment and utensils to be submersed in hot water maintained at 171°F (77°C) or above for at least 30 seconds. A properly calibrated thermometer is needed to routinely check the temperature of the sanitizing water. Employees must use dish baskets or racks to lower equipment and utensils into the sanitizing water. Manual warewashing operations rarely employ this type of sanitizing due to concerns about employee safety and the large amount of energy required to keep the water hot.

In all instances, a food-contact surface must be thoroughly cleaned and rinsed to remove soil and detergent residues before it can be properly sanitized.

◆ Heat sanitizing in manual warewashing operations requires cleaned equipment and utensils to be submersed in hot water maintained at 171°F (77°C) or above for at least 30 seconds.

Heat has several advantages over chemical sanitizing agents because it:

◆ Can penetrate small cracks and crevices
◆ Is noncorrosive to metal surfaces
◆ Kills all types of microorganisms equally effectively
◆ Leaves no residue
◆ Is easily measurable.

Hot Water

Sanitizing with hot water is most commonly performed as part of mechanical warewashing operations. The *Food Code* requires the hot water used for these purposes to be between 180°F (82°C) and 194°F (90°C) when it leaves the final rinse spray nozzles. This assures the temperature of the water will be at 171°F (77°C) or above when it reaches the surfaces of the equipment and utensils being sanitized. The only exceptions to these temperature requirements are single-tank, stationary-rack, and single-temperature machines, where the final rinse water temperature must be at least 165°F (74°C) as it leaves the spray nozzles.

Steam

Steam is an excellent agent for treating pieces of food equipment. The equipment shown here is an example of a low pressure, high temperature steam/vapor cleaning system. When steam in a flow cabinet is used, it should be sufficient to achieve 171°F (77°C) for at least 15 minutes or 200°F (94°C) for at least 5 minutes.

Steam is sometimes produced in remote steam boilers. Most boiler systems will require the periodic addition of boiler water additives to prevent the buildup of scale and corrosion in the boiler system. When there is a chance steam produced in these boilers will come into direct or indirect contact with food, the chemicals added to boiler water must be approved for use as a food additive and labeled for that use as per Title 21 CFR Part 173.310—Boiler Water Additives.

Low-pressure, steam cleaning system

When using heat for sanitizing, the temperature at the surface of the equipment and utensils must reach at least 160°F (71°C) to assure proper destruction of disease-causing microorganisms. The temperature at the surface can be measured by using irreversible heat-sensitive labels or tapes attached to a clean dry dish and then run with a load of soiled dishes through a regular wash cycle. The labels change from white to black when the indicated temperature is reached.

Heat-sensitive label

T-Stick

Measuring the sanitizing temperature in dishwashing machines

Another device for measuring the temperature of hot water sanitizers is the "maximum registering" or "holding" thermometer. This type thermometer will continue to hold the highest temperature measured until it is reset.

Maximum holding thermometers continue to hold the highest temperature measured until it is reset.

Chemical Sanitizing

The chemical **sanitizers** most commonly used in food establishments are **chlorine, iodophors,** and **quaternary ammonium compounds (quats)**. Iodophors are iodine-containing sanitizers, and quats are ammonia salts used as chemical sanitizers.

There are two ways to sanitize surfaces using chemical compounds:

◆ Immerse a piece of equipment or a utensil into a sanitizing solution at the prescribed concentration

◆ Swab, brush, or pressure spray the sanitizing solution directly onto the surface of the equipment and utensils.

The effectiveness of chemical sanitizers depends on a number of factors such as:

◆ **Concentration of sanitizer** — in general, increasing the concentration of a chemical sanitizer will increase its ability to destroy microbes. However, more is not always better. Using sanitizers at very high concentrations can be toxic, corrosive and wasteful.

◆ **Temperature of solution** — the effectiveness of a sanitizer typically increases as the temperature of the sanitizer solution increases up to about 120°F (49°C). However, at very high temperatures, the potency of a sanitizer decreases because it evaporates into the air.

◆ **pH or acidity of solution** — most sanitizers don't work well when the pH of the sanitizer solution is above 10. Therefore, effectiveness of sanitizers is reduced when used in hard water and when soap and detergent is not rinsed off the surfaces of equipment and utensils before they are sanitized.

◆ **Time of exposure** — sufficient time must be allowed for sanitizers to destroy disease-causing microorganisms. The amount of exposure time depends on the size of the microbial populations and their susceptibility to the sanitizer.

The chart on the next page summarizes the requirements for sanitizing equipment and utensils using chlorine, iodophors, quaternary ammonium compounds (quats) and hot water.

KEY TERMS

Sanitizers - approved substances or methods to use when sanitizing.

Chlorine - a chemical sanitizing compound commonly used in food establishments.

Iodophors - Iodine-containing sanitizers commonly used in food establishments.

Quarternary ammonium compounds (quats) - ammonia salts used as chemical sanitizers.

Requirements for Sanitizing Equipment and Utensils

Concentration range of sanitizer solution as expressed in parts per million (ppm) or mg/L	pH 10.0 or less and minimum water temperature	pH 8.0 or less and minimum water temperature	Contact time
Chlorine 25-49	120°F (49°C)	120°F (49°C)	Greater than or equal to 10 seconds
Chlorine 50-99	100°F (38°C)	75°F (24°C)	Greater than or equal to 7 seconds
Chlorine 100	55°F (13°C)	55°F (13°C)	Greater than or equal to 10 seconds
Idophors Greater than or equal to 12.5 and Less than or equal to 25	pH less than or equal to 5.0 or per label and water temperature at 68°F (20°C) or above		
Quaternary ammonium per label	Slightly alkaline pH is preferred, water hardness must be 500 ppm or less or per manufacturer's label, and water temperature at 75°F (24°C) or above. The *Food Code* recommends NOT using quats above 200 parts per million (ppm) for immersion sanitizing of food-contact surfaces.		Greater than or equal to 30 seconds
Hot water sanitize 3-compartment sink with integral heating device	Water temperature must be maintained at 171°F (77°C) or above and equipment and utensils completely immersed in rack or basket.		

NOTE: Chemical sanitizers and other chemical antimicrobials applied to food-contact surfaces shall be listed in 40 CFR 180.940 Tolerance Exemptions for Active and Inert Ingredients for use in Antimicrobial Formulations.

Source: 2009 Food Code

A chemical test kit or test strips must be available so warewashing personnel can routinely check the strength of the sanitizing solution.

Chemical sanitizing solutions should be checked on a regular basis and replaced with a fresh solution when it becomes contaminated or falls below the minimum level recommended by the manufacturer.

More information about the chemical sanitizers that are commonly used in food establishments and the factors that affect the action of chemical sanitizers can be found in Appendix H.

Mechanical Warewashing

Mechanical warewashing is performed in a dishwashing machine that cleans and sanitizes multiuse equipment and utensils automatically. Dishwashing machines are designed to clean and sanitize equipment and utensils that have no electrical parts and will fit into the machine. When properly maintained and operated, mechanical warewashing is as effective at removing soil and disease-causing microorganisms as is manual warewashing.

> **Mechanical warewashing equipment is designed to automatically clean and sanitize.**

Mechanical warewashing process

Mechanical warewashing uses an 8-step process to clean and sanitize equipment and utensils.

1. Pre-scrape and pre-flush soiled equipment and pre-soak utensils to remove visible soil.

2. Rack equipment and utensils so wash and rinse waters will spray evenly on all surfaces and the equipment will freely drain.

3. Wash equipment and utensils in a detergent solution.

4. Rinse equipment and utensils in clean water at a temperature consistent with the type of dishwashing machine being used.

5. Sanitize equipment and utensils in a fresh, hot water sanitizing rinse between 180°F (82°C) and 194°F (90°C) for at least 30 seconds, except for a single-tank, stationary-rack, or single-temperature machine, where the final rinse may not be less than 165°F (74°C) for at least 30 seconds. Water temperature at the surface of equipment and utensils must be at least 160°F (71°C) to achieve proper sanitizing. The recommended final rinse temperature for low temperature chemical sanitizing dishwashing machines is 120°F (49°C) or less.

   The flow pressure of the hot water sanitizing rinse must be in the range indicated on the machine manufacturer's data plate and not less than 5 or more than 30 pounds per square inch (PSI). The reduced water pressure is required in order to assure the sanitizing water covers the entire surface of the items being sanitized.

6. Drain properly and air-dry equipment and utensils. Do not towel dry food-contact surfaces.

7. Store clean and sanitary items in a clean, dry area where they are protected from contamination.

8. Clean and maintain the machine to keep it in proper working condition.

The temperature of the wash water (and power rinse if applicable) is extremely important in providing effective sanitizing of equipment and utensils. Water temperatures, water pressure, and conveyor speed or cycle time must be in accordance with the dish machine "data plate" and manufacturer's instructions.

Single-tank dishwashing machine

> **Operators of dishwashing machines must be careful not to contaminate cleaned and sanitized equipment and utensils by touching them with soiled hands.**

WAREWASHING: MECHANICAL & MANUAL		Minimum wash temperature	Minimum sanitizing temperature
SPRAY TYPE WAREWASHERS: Single-Tank, Hot Water Sanitize	Stationary-rack, single temperature	165°F (74°C)	165°F (74°C) *
	Stationary-rack, dual temperature	150°F (66°C)	180°F (82°C) *
	Conveyor, dual temperature	160°F (71°C)	
Multitank Hot Water Sanitize	Conveyor, multi-temperature	150°F (66°C)	
Chemical sanitize	Any warewashing machine	120°F (49°C)	Sanitization levels as stated in the *Food Code* or per labeled manufacturer's instructions on the container*
3-compartment sink	Cleaning agent labeling may permit lower washing temperatures	110°F (43°C)	

*Chemical sanitization units installed after the adoption of the FDA 2001 *Food Code* must be equipped with an audible or visual "low level" sanitizer indicator to add more sanitizer.

Cleaned and sanitized equipment and utensils must be properly drained and air-dried before storage. Soiled, multiuse cloth towels can recontaminate cleaned and sanitized equipment and utensils.

Cleaned and sanitized equipment and utensils must be stored in such a way as to protect them from contamination. Cabinets and carts used to store cleaned and sanitized equipment and utensils may not be located in locker rooms and toilet rooms and under sewer lines that are not shielded to catch drips, leaking water lines, open stairwells, or under other sources of contamination.

Proper storage of cleaned and sanitized equipment and utensils

According to Food Code:

◆ Cleaned and sanitized equipment and utensils must be properly drained and air-dried before storage.

◆ The *Food Code* prohibits cloth drying of equipment and utensils, with the exception that air-dried utensils can be "polished" with cloths that are clean and dry.

◆ If drying agents are used in conjunction with sanitization they must contain components that are Generally Recognized as Safe (GRAS) for use in food, GRAS for the intended purpose, or have been approved for use as a drying agent under the relevant provisions contained in Title 21 of the *Code of Federal Regulations*.

Manual Warewashing

Manual warewashing typically uses a 3- or 4-compartment sink to clean and sanitize equipment and utensils. The manual process begins with scraping and flushing food residues off the surface of equipment and utensils. When necessary, items may be pre-soaked to remove food residues and other soil.

Manual warewashing				
Pre-scrape	**Compartment No. 1 - Wash**	**Compartment No. 2 - Rinse**	**Compartment No. 3 - Sanitize**	**Air-dry**
◆ Items may be pre-soaked to remove food residues and other soil.	◆ Detergent solution with hot water ◆ Pre-flushed and pre-scraped equipment and utensils are washed.	◆ Fresh warm potable water ◆ Soap and soil are rinsed off equipment and utensils.	◆ Fill with chemicals or hot water ◆ Sanitize the previously cleaned surfaces.	◆ Air-dry dishes before putting them into storage.

Compartment No. 1 - After pre-scraping, the equipment and utensils are washed in the first compartment of the sink with warm water and a cleaning agent. Washing removes visible food particles and grease. Dishwashing personnel should always use the correct amount of detergent based on the quantity of water used in the wash compartment of the sink. The temperature of the wash water must be maintained at not less than 110°F (43°C) unless the manufacturer of the cleaning agent specifies a different temperature.

Compartment No. 2 - The second compartment of the sink is used to rinse equipment and utensils in clean, warm water. Rinsing removes cleaning agents, soap film, remaining food particles, and abrasives that may interfere with the sanitizer agent. For best results, the rinse water must be kept clean and the temperature should be between 110°F (43°C) and 120°F (49°C).

Compartment No. 3 - The third compartment of the sink is used for sanitizing items with either hot water or chemical sanitizers.

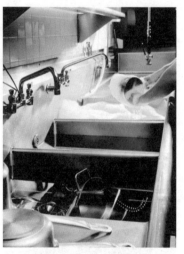

Manual warewashing

When using hot water:

◆ Temperature must be maintained at 171°F (77°C) or above at all times
◆ Items must be immersed in the hot water for at least 30 seconds.

When using chemical sanitizers:

◆ Preferred in food establishments because they do not require large amounts of hot water.
◆ The standard range of water temperatures for chemical sanitizing solutions is between 68°F (20°C) and 120°F (49°C).
◆ Under certain conditions chlorine sanitizers may be used with a water temperature as low as 55°F (13°C).
◆ Quats may be used with a water temperature above 120°F (49°C). Follow the manufacturer's directions for proper use and concentration of quats.

Hot and cold potable water must be supplied to each compartment of the sink, and the sink should be cleaned and sanitized prior to use.

◆ Chlorine and iodine sanitizers should not be used in water temperatures above 120°F (49°C) because the potency of these sanitizers is quickly lost at high water temperatures.

The manual warewashing sink must be equipped with sloped drain boards or dish tables of adequate size to store soiled items prior to washing, and clean items after they have been sanitized.

As with mechanical warewashing, items manually cleaned and sanitized must be air-dried before they are put into storage. If drying agents are used in conjunction with sanitization they must contain components that are GRAS for the intended purpose or have been approved for use as a drying agent under the relevant provisions contained in Title 21 of the *Code of Federal Regulations.*

The effectiveness of the manual warewashing process depends on the condition of the wash, rinse, and sanitizing solutions. Monitor wash and rinse solutions and replace them when they become soiled. When the sanitizer in the third compartment is depleted, it should be drained completely and replaced with a fresh solution at proper strength. The concentration of a chemical sanitizer solution must be tested periodically with a test kit or test strips to make sure it remains at strength. These strips provide a color comparison to indicate the strength of the sanitizer. Test kits or strips may be obtained from the manufacturer of your sanitizer.

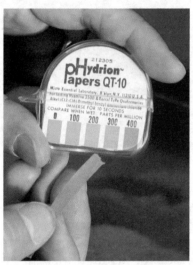

Measuring the concentration of sanitizers using test strips

Cleaning Fixed Equipment

Cleaning fixed equipment

Fixed equipment used in food establishments has food-contact surfaces but cannot be cleaned using either traditional mechanical or manual warewashing processes. Common examples of fixed equipment include band saws, floor-mounted mixers, slicers, grinders, and live tanks used to display live lobsters and molluscan shellfish. This equipment must be disassembled to expose food-contact surfaces to cleaning and sanitizing agents. The basic steps for cleaning fixed equipment are listed on the next page. Large equipment such as preparation tables can be cleaned by using a foam or spray method. In this process, detergents and degreasers, fresh water rinse, and a chemical sanitizer are applied using foam or spray guns. The hoses, feed lines, and nozzles that make up the foam or spray unit should be in good condition and attached properly.

The bucket and brush method is often used to clean equipment that could be damaged by pressure spraying or immersion in the manual warewashing sink. This system uses one or more buckets for washing, rinsing, and sanitizing.

Steps for cleaning fixed equipment

1. Disconnect the power supply to the equipment before taking it apart for cleaning.

2. Disassemble the equipment as necessary to allow the detergent solution to reach all food-contact surfaces.

3. Use a plastic scraper to clean equipment parts and remove food debris that has accumulated under and around the equipment. Scrape all debris into the trash.

4. Carry the parts that have been removed from the equipment to the manual warewashing sink where they will be washed, rinsed, and sanitized.

5. Wash the remaining parts of the equipment using a clean cloth, brush, or scouring pad and warm, soapy water. Clean from top to bottom.

6. Rinse thoroughly with fresh water and a clean cloth.

7. Swab or spray a chemical sanitizing solution, mixed to the manufacturer's recommendations, onto all food-contact surfaces.

8. Allow all parts to drain and air-dry.

9. Reassemble the equipment.

10. Resanitize any food-contact surface that might have been contaminated due to handling when the equipment was being reassembled.

Wash removable parts in a manual warewashing sink

Wash remaining parts with a clean cloth and warm soapy water.

Steps in cleaning fixed equipment

Food-contact surfaces have to be cleaned and sanitized while non–food–contact surfaces require cleaning only.

The clean-in-place (CIP) method is used for equipment that is designed to be cleaned and sanitized by circulating cleaning and sanitizing solutions through the equipment. Examples of equipment that would be cleaned using this method are soft-serve ice cream or yogurt machines. The basic steps in the clean-in-place method are listed in the following table.

Clean-in-place cleaning method

Soft-serve ice cream machine

(Courtesy of SaniServ)

1. Empty food product and waste from the equipment.

2. Disconnect the power to the equipment.

3. Disassemble if parts are removable and cleanable in the manual warewashing sink.

4. Clean and sanitize the removable parts using the manual warewashing process.

5. Clean and sanitize the main unit by circulating wash, rinse, and sanitizing solutions through it. Combination detergent sanitizers can also be used on equipment that is designed for CIP cleaning.

6. Reassemble parts removed from the main unit.

7. Resanitize by circulating a manufacturer approved sanitizing solution through the equipment.

Wiping cloths

Establishments will use wiping cloths to clean up spills. These cloths should be stored in a container of sanitizing solution between uses in order to prevent microbial growth on the cloth. This is often referred to as an "in-place sanitizer." Any standard chemical sanitizer may be used for storing wiping cloths. However, quaternary ammonia compounds are preferred because they kill disease-causing microorganisms for a longer period of time. Some important rules to follow when using wiping cloths include:

◆ A clean cloth and fresh container of sanitizing solution at the proper strength must be used at the start of each day
◆ The cloth should be changed whenever it becomes heavily soiled and the sanitizer must be changed when it falls below the required concentration
◆ Wiping cloths may not be used for any other purpose except removing spills
◆ Cloths used to wipe spills from floors and non-food-contact surfaces must be kept separate from the cloths used to clean food-contact surfaces of equipment

Store and use wiping cloths properly.

◆ Cloths used for wiping floors and surfaces that have come in contact with raw animal foods must be used for no other purpose.

A food establishment should provide single-use and single-service articles for food employees if it does not have proper facilities for cleaning and sanitizing multi-use kitchenware and tableware. These establishments must also provide single-service articles for use by customers.

Materials used to make single-service and single-use articles must not permit the transfer of harmful substances or pass on colors, odors, or tastes to food. These materials must be safe and clean when used in food establishments.

Single-use items must be stored and dispensed in a way that will protect them from contamination by employees, customers, and the surrounding environment.

Cleaning Environmental Areas

Cleaning environmental areas

A regular cleaning schedule of non-food-contact surfaces should be established and followed to maintain the facility in a clean and sanitary condition. The manager or supervisor should create a master cleaning schedule that will list the following items:

◆ The specific equipment and facilities to be cleaned

◆ The processes and supplies needed to clean the equipment and facilities

◆ The prescribed time when the equipment and facilities should be cleaned

◆ The name of the employee who has been assigned to do the cleaning.

> **Major cleaning should be done during periods when the least amount of food will be exposed to contamination and service will not be interrupted.**

Ceilings and Walls

Ceilings should be checked regularly to make certain they are not contaminating food production areas. Ceilings, lights, fans, and covers can be cleaned using either a wet- or dry-cleaning technique. When wet-cleaning ceilings and fixtures, it is best to use a bucket method to keep water away from lights, fans, and other electrical devices. Walls may be cleaned using either the bucket or spray methods. Whenever possible, disconnect power before cleaning fixtures.

Floors

The cleanliness of floors depends on the ability of an employee to remove soil from the surface of the floor. Floors in food production areas can be cleaned using a spray system for washing and rinsing. Floors that will be damaged by spray cleaning can be cleaned using the bucket method.

If floor drains are not properly maintained, they can be a source of disease, vermin, and odors. All floor drains should be flushed, washed, rinsed, and sanitized. The steps commonly used to clean floor drains are listed below:

1. Remove the grate or cover over the drain.

> **All floors should be sloped to drains to help remove soil and the water used to clean floors.**

2. Clean out waste and other debris from the drain.

3. Use a sprayer or hose to flush the drain and grate or cover.

4. Pour in drain cleaner to break up grease and other waste in the drain.

5. Wash the drain using a brush or water pressure from the sprayer.

6. Rinse the drain with hot water.

7. Pour or spray sanitizer into the drain.

If not properly maintained, sewer pipes can become clogged and can cause sewage and other wastes to back up into and flood various areas of a food establishment. This can create both a health hazard and a nuisance. You must know what to do in the event a sewer backup occurs. Some of the steps you should follow are:

Floor drains should be flushed, washed, rinsed, and sanitized.

◆ Remove anything from the floor that might be damaged or contaminated by the wastewater.

◆ Remove the blockage in the pipe causing the backup. This may require the services of a professional plumber.

◆ Wash and sanitize all equipment, floors, walls, and any other objects that may have been contaminated by the waste that backed up into the establishment.

Equipment and Supplies Used for Cleaning

Some examples of equipment commonly used during cleaning are nylon brushes, cleaning cloths, scouring pads, squeegees, mops, buckets, spray bottles, hoses, and spray or foam guns. Commonly used cleaning supplies include hot water, cleaners, degreasers, and sanitizers.

There should be a separate sink to fill and empty mop buckets, to rinse and clean mops, and to clean brushes and sponges. A "janitor's" sink or floor drain should be provided to dispose of wastewater produced by cleaning activities.

Janitor sink

Containers of poisonous or toxic materials and personal care items must bear a legible manufacturer's label. Working containers used for storing poisonous or toxic materials such as cleaners and sanitizers taken from bulk supplies must be clearly and individually identified with the common name of the material. A container previously used to store poisonous or toxic materials may not be used to store, transport, or dispense food.

The Occupational Safety and Health Administration (OSHA) requires employees to have the "Right-to-Know" about the chemicals to which they may be exposed on the job. Information about chemical substances in the workplace is provided to employees by way of Material Safety Data Sheets (MSDS). Chemical manufacturers develop these and they must be maintained on file and accessible to employees in the food establishment.

Examples of information contained in an MSDS include:

◆ Ingredients

◆ Physical and chemical characteristics

◆ Fire, explosion, reactivity, and health hazard data

◆ How to handle chemicals safely

◆ How to use personal protective equipment and other devices to reduce risk

◆ Emergency procedures to use if required.

Failure to produce the required MSDS upon demand can result in a substantial fine from OSHA.

MATERIAL SAFETY DATA SHEET

PRODUCT IDENTIFICATION
Product:

    Product Code:

Chemical Name/Synonym: Oxypurinol Xanthine

INGREDIENTS
Oxypurinol	1%	CAS#:2465-59-0
Xanthine	1%	CAS#:69-89-6

PHYSICAL AND CHEMICAL CHARACTERISTICS
Boiling Point: NA

Melting Point: NA

Vapor Pressure (mm Hg): NA

Vapor Density (air=1): NA

Solubility in Water: Insoluble

pH: NA

Specific Gravity: NA

Bulk Density: Not Determined

Evaporation rate: NA

Percent Volatile: NA

Appearance and Odor: Solid, slightly musty

FIRE AND HAZARD DATA
Flash Point: > 220°F

Flammable Limits: NA

Autoignition Temp: NA

Decomposition Temp: NA

NTP Teratogen or Mutagen Carcinogen: Not Determined

IARC or OSHA Potential Carcinogen: Not Determined

Fire Extinguishing Media: Carbon Dioxide, dry chemical or foam.

Special Fire Fighting Procedures: Wear self-contained breathing apparatus and protective clothing to prevent contact with skin and eyes.

Unusual Fire and Explosion Hazards: Emits toxic fumes under fire conditions.

HEALTH HAZARDS
Primary Route(s) of Entry: Skin, eye, inhalation, ingestion.

Signs of Symptoms of Exposure: May cause irritation. May cause allergic skin reaction.

Target Organs: Liver. Complete chemical, physical and toxicological properties have not been thoroughly investigated.

EMERGENCY FIRST-AID PROCEDURES
Ingestion: If swallowed, wash out mouth with water provided person is conscious. Call a physician.

Skin: Flush with copious amounts of water for at least 15 minutes. Remove contaminated clothing and shoes. Call a physician.

Eyes: Flush with copious amounts of water for at least 15 minutes. Assure adequate flushing by separating the eyelids with fingers. Call a physician.

Inhalation: Remove victim to fresh air. If breathing is difficult, call a physician.

Note to Physician: No specific antidote is available. Treatment of overexposure should be directed at the control of symptoms and the clinical conditions.

TOXICITY INFORMATION:
Oral LD$_{50}$ (Rats): > 5,000 mg/kg

Dermal LD$_{50}$ (Rabbits): >2,000 mg/kg (Technical)

Eye Irritation: Slight

Skin Irritation: Very slight

REACTIVITY
Stability: Stable

Conditions to Avoid: None

Incompatibility: Strong oxidizing agents

Hazardous Decomposition Products: Thermal decomposition may produce carbon monoxide, carbon dioxide, and nitrogen oxides.

SPILL, LEAK, AND DISPOSAL PROCEDURES
If material is spilled: Sweep, vacuum or shovel material into a container for reuse or disposal. Do not allow product to contaminate drains, sewers, streams, ditches or bodies of water. Prevent large quantities from contacting vegetation. Keep animals away from large spills. Dike and contain spill area, then rinse area and tools several times with soapy water. Dispose of rinse water in accordance with applicable law.

Waste Disposal: Dispose according to Federal EPA procedures as outlined in the Resource Conservation Recovery Act (RCRA) and follow state and local guidelines. Empty containers may contain some product residues and should be handled and disposed of in a similar manner to the product. All attempts should be made to utilize the product completely, in accordance with its intended use.

STORAGE AND HANDLING
Precautions: Store in a cool, dry place. Separate from other pesticides, fertilizers, seed, feed, foodstuffs and away from drains, sewers and water sources.

Other Precautions: Keep out of reach of children. Practice good care and good safety precautions when handling this product. Avoid contact with eye, skin and clothing. Avoid breathing dust. Do not swallow. Wash thoroughly after handling. This product is toxic to fish. Do not apply directly to lakes, ponds or streams. This product may be an attractant to pets and rodents. Store in a secure place.

FEDERAL REGULATORY INFORMATION:
SARA TITLE III; SEC. 311/312 Hazard Categories

Immediate (Acute) Health: NA

Delayed (Chronic) Health: NA

Fire: NA

Sudden Release of Pressure: NA

Reactivity: NA

SEC 302/SEC 304 TPQ: Not Listed

SEC 313: NA

CERCLA RQ: NA

CAA RQ: NA

EPA Reg. No.: 1001-73

HM- 181 Shipping Name: Insecticide, Agricultural, Solid, NOI

NA = Not Applicable
ND = Not Determined

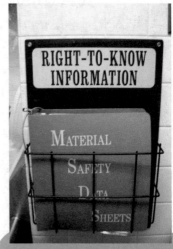

Keep material safety data sheets together for easy access.

Material safety data sheet

Summary

What Do You Think?

Broken hot water heater contributes to Hepatitis A virus outbreak at family dining restaurant

Approximately 12 people became ill with a Hepatitis A virus infection at a family dining restaurant. Health officials suspect the outbreak was caused by an infected restaurant employee who contaminated food items and equipment during food preparation. The problem was compounded by the fact that the water heater in the establishment was broken. According to employees of the restaurant, hot water was not available to wash dishes, clean tables, or wash hands for several days.

What problems occurred and what would you do in this situation?

Back to the Story... The case presented at the beginning of the chapter illustrates how the risk of foodborne illness can be increased significantly in food establishments that do not have an adequate supply of hot water. Many foodborne diseases can be transmitted very easily by employees who do not follow good personal hygiene practices. This includes employees who do not wash their hands after using the toilet. Pathogens can also be spread by contaminated equipment and utensils. An adequate supply of hot water is vital for washing hands and cleaning and sanitizing equipment and utensils. Failure to provide an adequate supply of hot water is a critical violation that should be corrected immediately by the person in charge of a food establishment.

A good sanitation program starts with a neat, clean, and properly maintained building. A well-maintained facility helps protect food products from contamination, makes proper stock rotation easier, prevents entrance of pests, reduces fire hazards, and contributes to the overall safety of the work environment. Proper cleaning and sanitizing also enhance the safety and quality of food and increase the life expectancy of equipment and facilities.

A food establishment's sanitation program consists of the standards, policies, and procedures that assure proper cleaning and sanitation. In-store sanitation standards must incorporate the requirements of government regulations and manufacturers' guidelines.

The effectiveness of a food establishment's sanitation program is directly related to the degree of commitment and concern demonstrated by top management. Managers must motivate their employees to follow prescribed sanitation practices and reward those employees who do so successfully.

Proper education and training is required to assure proper sanitation practices are followed. Employees responsible for handling food, equipment, and cleanup of the building and grounds must be trained to accept sanitation as one of their key responsibilities. A clean, well-maintained facility will become a source of pride for employees.

For a sanitation program to be effective, cleaning tasks must be scheduled and individual employees assigned to complete the tasks. Employees must be given instructions that include what they are to clean, how to clean it, and what tools and supplies are required to effectively clean it. Properly trained employees must carry out cleaning activities, and they must be closely monitored by supervisors to identify and correct problems when they occur.

A good sanitation program is a preventive program that anticipates and eliminates potential hazards before they become serious problems. The value of an effective program is difficult to measure in terms of dollars. Nonetheless, the value of effective sanitation can be evident in the form of diminished problems and increased goodwill with customers and the local regulatory community.

Quiz 6 (Multiple Choice)

Please choose the BEST answer to the questions.

1. Which of the following procedures must be performed daily in a food establishment?

    a. Clean and sanitize floors, walls and ceilings.

    b. Wash and sanitize drink dispensers.

    c. Clean shelves in the dry storage area.

    d. Clean and sanitize soiled equipment and utensils.

2. What are the proper steps in manual and mechanical warewashing operations after pre-scraping?

    a. Wash, rinse, sanitize, and towel dry.

    b. Rinse, wash, sanitize, and air-dry.

    c. Wash, rinse, sanitize, and air-dry.

    d. Rinse, wash, sanitize, and towel dry.

3. When sanitizing with hot water in a manual warewashing operation, the temperature of the water in the final rinse must be maintained at:

    a. 161°F (72°C).

    b. 171°F (77°C).

    c. 181° F (83°C).

    d. 191°F (88°C).

4. Which of the following statements is **false**?

    a. Dishes should be washed in very hot water (above 171°F, 77°C) to effectively remove soil from the surface.

    b. Pre-scraping helps to remove larger food particles from dishes, which helps keep the wash water clean.

    c. The cleaning compounds used in a food establishment must be tailored to the individual water supply.

    d. Cleaning is a process which removes soil and prevents accumulation of food residues on equipment, utensils, and surfaces.

5. The strength of the chemical sanitizer in the third compartment of the 3-compartment sink must be checked frequently because:

    a. If the chemical is too strong, it may ruin fragile dishware.

    b. The chemical strength increases over time and may leave a toxic residue on the surface of equipment and utensils.

    c. The strength of chemical sanitizers may drop off as germs are killed off and the sanitizer is diluted with rinse water.

    d. The strength of the chemical increases as germs are killed off.

6. Which of the following statements about chemical sanitizing is (false?)

   a. Chlorine sanitizers may be used in water temperatures between 55°F (13°C) and 120°F (49°C).

   b. Iodine sanitizers may be used in water temperatures between 68°F (20°C) and 120°F (49°C).

   c. Quaternary ammonium (quat) sanitizers may be used in water temperatures above 120°F (49°C).

   (d.) Most chemical sanitizers work best when used with very hot water.

7. The process used to remove soil from the surface of equipment and utensils is:

   a. Cleaning.

   b. Sanitizing

   c. Pasteurization.

   d. Irradiation.

8. The process used to reduce the number of disease-causing microorganisms from the surfaces of equipment and utensils is:

   a. Cleaning.

   b. Sanitizing

   c. Pasteurization.

   d. Irradiation.

9. The first step in developing a cleaning program for environmental areas in a food establishment is to:

   a. Determine the cleaning needs.

   b. Assign cleaning jobs to employees.

   c. Obtain the equipment and supplies needed to put the cleaning program into operation.

   d. Conduct a training program to teach employees how to clean properly.

10. Which of the following pieces of information is **not** provided by a material safety data sheet?

   a. The physical and chemical characteristics of a hazardous substance used in the establishment.

   b. How to use personal protective equipment and other devices to reduce an employee's risk of injury.

   c. Emergency procedures, such as first aid, to use when exposed to a hazardous chemical.

   (d.) A list of approved sites for disposing of hazardous materials.

Answers to the multiple-choice questions are provided in **Appendix A**.

References/Suggested Readings

Code of Federal Regulations. 2010. 21 CFR 173.310 - Boiler Water Additives. U.S. Government Printing Office, Washington, D.C.

Code of Federal Regulations. 2001. 21 CFR 172 - Substances Generally Recognized as Safe. U.S. Government Printing Office, Washington, D.C.

Code of Federal Regulations. 2010 21 CFR 184 - Direct Food Substances Affirmed as Generally Recognized as Safe. U.S. Government Printing Office, Washington, D.C.

Code of Federal Regulations. 2010. 21 CFR 186 - Indirect Food Substances Affirmed as Generally Recognized as Safe. U.S. Government Printing Office, Washington, D.C.

Food and Drug Administration. 2009. *2009 Food Code.* U.S. Public Health Service, Washington, D.C.

Suggested Web Sites

All QA Products
  www.allqa.com

Bowerman Marketing Group
  www.ebowerman.com

Champion Industries
  www.championindustries.com

Chemstar
  www.chemstarcorp.com

Cooper-Atkins Corp.
  www.cooper-atkins.com

DeltaTRAK, Inc
  www.deltatrak.com

Duke Manufacturing Co.
  www.dukemfg.com

Ecolab
  www.ecolab.com

Hobart Corp.
  www.hobartcorp.com

Johnson Diversey
  www.johnsondiversey.com

Paper Thermometer Co.
  www.mv.com/ipusers/paperthermometer

The Soap and Detergent Association
  www.sdahq.org

The Steritech Group, Inc.
  www.steritech.com

NSF International
  www.nsf.org

Underwriters Laboratories, Inc.
  www.ul.com

T-Stick.com

   www.t-stick.com

The Food Marketing Institute

   www.fmi.org

Gateway to Government Food Safety Information

   www.foodsafety.gov

The National Restaurant Association

   www.restaurant.org

Notes

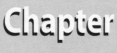

Chapter 7

Environmental Sanitation and Maintenance

Key Terms

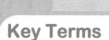

- ◆ Air gap
- ◆ Backflow
- ◆ Back pressure
- ◆ Back siphonage
- ◆ Coving
- ◆ Cross connection
- ◆ Garbage
- ◆ Integrated Pest Management (IPM)
- ◆ Refuse
- ◆ Vacuum breaker

The appearance of a food establishment, including the building and grounds, will influence a customer's overall dining or shopping experience. Customers are most favorably impressed by a building that is neat and clean and restrooms that are well maintained. Proper storage and disposal of refuse and garbage are important to control foul odors and discourage pests that can damage food and contaminate food-contact surfaces.

Learning Points:
- Describe how the visual appearance of a food establishment can affect a guest's opinion about the cleanliness and sanitation of the operation.
- Identify the equipment and supplies that must be provided in a properly equipped restroom facility.
- Identify the main components of a properly equipped handwashing sink.
- Explain how conveniently located handwashing facilities contribute to good personal hygiene.
- Identify the types of plumbing hazards that can have a negative effect on public health.
- Identify some types of pests and common signs of pest infestation that may be found in food establishments.
- Describe how integrated pest management is used to control pests in food establishments.
- Explain how proper disposal and storage of garbage and refuse help prevent contamination and pest problems in a food establishment.

Condition of the Food Establishment

The exterior of a food establishment must be free of litter and debris that could attract and harbor pests and ruin the appearance of the facility. Grass and weeds should be regularly mowed to eliminate harborage areas for insects, rodents, and other pests. Walking and driving surfaces should be constructed of concrete, asphalt, gravel, or similar materials to facilitate maintenance and control dust. These surfaces should also be properly graded to prevent rainwater from pooling and standing on parking lots and sidewalks. To attract customers, the exterior of the building should be clean, attractive and make a good impression.

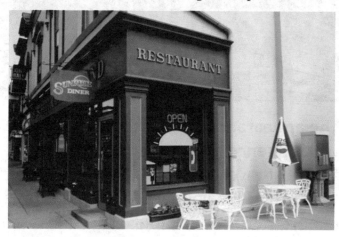

Proper Water Supply and Sewage Disposal System

An adequate water supply and proper sewage disposal are vital to the sanitation of food establishments. The capacity of the water source and the ability to produce hot water should be sufficient to meet the demands of the food establishment. Drinking water for food establishments must be obtained from an approved source. Most establishments will be connected to a public water system. However, when a private well or other nonpublic water system is used, it must be constructed, maintained, and operated according to the water quality requirements of the jurisdiction. Wells must be located and constructed in a manner that will protect them from sewage and other sources of contamination. Periodic sampling is required to monitor the safety of the water and to detect any change in water quality.

A reservoir used to supply water to devices such as a fountain beverage dispenser, drinking water vending machine, or produce mister must be maintained in accordance with the manufacturer's specifications. The reservoir must be cleaned at least once a week in accordance with manufacturer's specifications, or according to the following procedures, whichever is more stringent:

1. Drain and completely disassemble the water and aerosol contact parts.
2. Brush-clean the reservoir, aerosol tubing, and discharge nozzles with a suitable detergent solution.
3. Flush the complete system with water to remove the detergent solution and particle accumulation.
4. Rinse by immersing, spraying, or swabbing the reservoir, aerosol tubing, and discharge nozzles with at least 50 ppm hypochlorite solution.

Proper disposal of sewage greatly reduces the risk of fecal contamination of food and water. The *Food Code* requires sewage from food establishments to be disposed through an approved facility that is:

◆ A public sewage treatment plant, or
◆ An individual sewage disposal system that is sized, constructed, maintained, and operated according to rules and regulations of the jurisdiction.

The use of nonpotable water sources in retail food establishments must be approved by the regulatory authority in the jurisdiction and may only be used for nonculinary purposes, such as air handling systems, cooling systems, and fire protection.

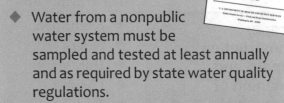

According to Food Code:

◆ Water from a nonpublic water system must be sampled and tested at least annually and as required by state water quality regulations.

◆ To render the water safe, a drinking water system must be flushed and disinfected before being placed into service. The *Food Code* prescribes that a drinking water system is flushed after construction, repair, or modification and after an emergency situation, such as flood, that may introduce contaminants to the system.

Condition of Building

The cleanliness and attractiveness customers view upon entering the building influence their overall dining or shopping experience. Entrance doors should be self-closing to discourage flying insects from entering the building.

A pleasant facility attracts customers.

The entrance area to a food establishment creates a lasting impression.

Floors, Walls, and Ceilings

Consider the specific needs of the different food areas when selecting materials for floors, walls and ceilings. Some criteria that should be used for all departments are:

◆ Sanitation
◆ Safety
◆ Durability
◆ Comfort
◆ Cost.

Materials used for floors, walls, and ceilings in food preparation areas, store rooms (including dry storage areas and walk-in refrigerators), dishwashing areas, and restrooms must be:

◆ Smooth
◆ Nonabsorbent
◆ Easy to clean
◆ Resistant to damage and deterioration.

> **Proper construction, repair, and cleaning of floors, walls, and ceilings are important elements of an effective sanitation program.**

Floors

Floors graded to drains are needed in food establishments where water flush methods are used for cleaning. In addition, the floor and wall must be coved and sealed. **Coving** is a curved sealed edge between the floor and the wall that eliminates sharp corners or gaps that would make cleaning difficult and ineffective. When cleaning methods other than water flushing are used for cleaning floors, the floor and wall juncture must be coved with a gap of no more than 1/32 inch (1 mm) between the floor and wall.

Use mats and other forms of antislip floor coverings where necessary to protect employees from slips and falls. These devices should also be impervious, nonabsorbent, and easy to clean.

> **According to Food Code:**
>
> The *Food Code* prohibits the use of carpeting in:
> ◆ Food preparation areas
> ◆ Walk-in refrigerators
> ◆ Warewashing areas
> ◆ Toilet room areas where handwashing sinks, toilets, and urinals are located
> ◆ Refuse storage rooms or other areas subject to moisture.

KEY TERM

Coving - a curved sealed edge between the floor and wall that eliminates sharp corners or gaps that would make cleaning difficult and ineffective.

Coving

Anti-slip mats help prevent falls.

Walls and Ceilings

Walls and ceilings in food production and warewashing areas must be made of a light colored material to enhance the artificial lighting in these areas. This will make soil and dirt easier to see and will help employees know when they have done an effective job cleaning a surface. Walls and wall coverings should be constructed of materials such as ceramic tile, stainless steel, or fiberglass when used in areas cleaned frequently.

Ceilings should be constructed of nonporous, easily cleanable materials. Studs, rafters, joists, or pipes must not be exposed in walk-in refrigeration units, food preparation and dishwashing areas, and toilet rooms.

Light fixtures, ventilation system components, and other attachments to walls and ceilings must be easy to clean and maintained in good repair.

Restroom Sanitation

Toilet facilities near work areas promote good personal hygiene, reduce lost productivity, and permit closer supervision of employees. Toilet rooms in food establishments must be completely enclosed and provided with tight-fitting and self-closing doors.

Materials used in the construction of toilet rooms and toilet fixtures must be durable and easily cleanable. The floors, walls, and fixtures in toilet areas must be clean and well maintained. Supply toilet tissue at each toilet. Provide easy-to-clean containers for waste materials and have at least one covered container in toilet rooms used by women.

Employee restrooms must be conveniently located and accessible to employees during all hours of operation.

Poor sanitation in toilet areas can spread disease.

Never store food or items associated with food handling in restroom areas.

Handwashing Facilities

Food employees must know when and how to wash their hands. Refer to Chapter 3 for details on the proper way to wash your hands. Conveniently located and properly equipped handwashing facilities are key factors in getting employees to wash their hands. Handwashing sinks should be located in or near areas where food is prepared or handled and in warewashing areas. Handwashing sinks must also be located in or adjacent to restrooms. The number of handwashing sinks required and their installation are usually set by local health or plumbing codes.

A handwashing sink must be equipped with:

◆ Hot and cold running water under pressure

◆ A supply of soap

◆ A way to dry hands without contaminating them (paper towels or air dryer)

◆ A waste receptacle for paper towels

◆ A handwashing sign or poster that notifies employees to wash their hands must be clearly visible at all handwashing sinks used by food employees.

Handwashing stations should be well maintained and never used for purposes other than hand washing.

Don't use produce sinks for hand washing.

Water flow and temperature - A handwashing sink must be able to provide water at a temperature of at least 100°F (38°C) through a mixing valve or combination faucet. If a self-closing, slow-closing, or metering faucet is used, it must provide a flow of water for at least 15 seconds without the need to be reactivated.

Soap - Each handwashing sink must be equipped with a dispenser containing liquid or powdered soap. The use of bar soap is frequently discouraged by regulatory agencies because bar soap can become contaminated with germs and soil.

Hand drying - Individual disposable towels and mechanical air dryers are the preferred hand drying devices. Most local health departments do not allow retractable cloth towel dispenser systems because there are too many possibilities for contamination. Common cloth towels, used multiple times by employees to dry their hands, are also prohibited.

Plumbing Hazards in Food Establishments

A properly designed and installed plumbing system is very important to food sanitation. The *Food Code* includes many different components within the definition of a plumbing system, such as:

◆ Water supply and distribution pipes
◆ Plumbing fixtures and traps
◆ Soil, waste, and vent pipes
◆ Sanitary and storm sewers
◆ Building drains, including their respective connections and devices within the building and at the site.

The plumbing system in a food establishment must be maintained in good repair. When repairs to the plumbing system are required, they must be completed in accordance with applicable local ordinances and state regulations.

Cross Connections and Backflow

A **cross connection** may be either direct or indirect. A direct cross connection occurs when a potable water system is directly connected to a drain, sewer, nonpotable water supply, or other source of contamination. An indirect cross connection is where the source of contamination (sewage, chemicals, etc.) may be blown across, sucked into, or diverted into a safe water supply.

Backflow occurs most frequently under two conditions:

◆ **Backpressure** where contamination is forced into a potable water system through a connection that has a higher pressure than the water system.

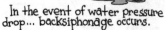
In the event of water pressure drop... backsiphonage occurs.

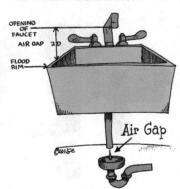
Air Gap

◆ **Backsiphonage**
Occurs when there is reduced pressure or a vacuum formed in the water system. This might be caused by a water main break, the shutdown of a portion of the system for repairs, or heavy water use during a fire.

Methods and Devices to Prevent Backflow

The plumbing system in a food establishment must be designed, constructed, and installed according to the plumbing code in the local jurisdiction. A properly designed and installed plumbing system will keep food, equipment, and utensils from becoming contaminated with disease-causing microorganisms found in sewage and other pollutants.

KEY TERMS

Cross connection - any physical link through which contaminants from drains, sewers, or waste pipes can enter a potable (safe to drink) water supply.

Backflow - the backward flow of contaminated water into a potable water supply.

Devices to prevent cross connections and backflow

KEY TERMS

Air gap - an unobstructed, open vertical distance through air that separates an outlet of the potable water supply from a potentially contaminated source like a drain.

Vacuum breaker - designed for use under a continuous supply of pressure. Spring loaded device to operate after extended periods of hydrostatic pressure.

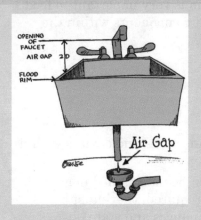

Air gap

- ◆ The most dependable backflow prevention device
- ◆ Vertical air space separates potable and nonpotable systems
- ◆ Distance of air space must be at least 2 times the diameter of the supply pipe (2D), but never less than 1 inch (25 mm).

Atmospheric vacuum breaker

- ◆ Atmospheric vent used in combination with a check valve
- ◆ Valve's air inlet closes when the potable water flows in the normal direction, but as water ceases to flow the air inlet opens, thus interrupting the possible backsiphonage effect.

Pressure type vacuum breaker

- ◆ Designed for use under continuous supply pressure, but not effective under backpressure
- ◆ Use only where nonpressure vacuum breakers cannot be used.

Dual check valve

- ◆ Two internally loaded, independently operating check valves that may be used as protection for all direct connections through which foreign materials might enter a potable water system.

Reduced pressure principle backflow preventers

- ◆ Used on all direct connections that may be subject to backpressure or backsiphonage
- ◆ Used when there is a possibility of contamination by material that is a health hazard.

Backflow Prevention Devices on Carbonators

Carbonators on soft drink dispensers form carbonic acid by mixing carbon dioxide with water. The carbonic acid is then mixed with the syrups to produce the soft drinks. If the carbon dioxide backs up into a copper water line, the carbonic acid will dissolve some of the copper. The water containing the dissolved copper will then be used in dispensing soft drinks, and the first few customers receiving the drinks are likely to suffer the symptoms of copper poisoning. An air gap or a vented backflow prevention device meeting American Society of Sanitary Engineering (ASSE) Standard No. 1022 must be installed upstream from a carbonating device and downstream from any copper in the water supply line to reduce incidences of copper poisoning.

Backflow preventer on a carbonator

Grease Traps

Food establishments that do a lot of frying or charbroiling should be equipped with a grease trap. These devices remove liquid grease and fats after they have hardened and become separated from the wastewater. Grease traps are especially important when the food establishment is connected to a septic system or other type of on-site wastewater treatment and disposal system. A grease trap must be located for easily accessible cleaning.

Grease trap

> Failure to properly clean and maintain grease traps can result in the harborage of pests and/or the failure of the facility's sewage system.

Garbage and Refuse Sanitation

Proper storage and disposal of **garbage** and **refuse** are required at food establishments to protect food and equipment from contamination. Insects, rodents, and other pests are also less likely to be attracted to these establishments when garbage and refuse are properly managed. Effective waste management requires:

◆ Proper handling and short-term storage of the materials inside the operation
◆ Proper storage of the waste outside the building until it is picked up by a commercial refuse disposal company.

KEY TERMS

Garbage - food wastes that cannot be recycled.

Refuse - trash, rubbish, and other types of solid waste not disposed of through the sewage system.

Inside Storage

Waste containers must be provided in all areas in a food establishment where refuse is produced or discarded. Containers used to collect garbage and refuse must be:

◆ Durable
◆ Cleanable
◆ Insect and rodent-proof
◆ Leak-proof
◆ Nonabsorbent
◆ Covered with a tight-fitting lid when not in use.

Plastic bags and wet strength paper bags are frequently used to line waste containers. Do not place waste containers in locations where they might create a public health nuisance.

Refuse storage rooms and containers must be cleaned as part of the establishment's routine cleaning program. Keeping the area clean is your best defense against pests. When cleaning this equipment, be careful not to contaminate food, equipment, utensils, linens, or single-service and single-use articles. Wastewater produced while cleaning the equipment and receptacles is considered to be sewage. It must be disposed of through an approved sanitary sewage system or other system constructed, maintained, and operated according to law.

Waste storage container with plastic liner

Outside Storage

A food establishment should also have an outside storage area and enclosure to hold refuse, recyclables, and returnables awaiting pickup. An outdoor storage surface should be durable, cleanable, and maintained in good repair. Dumpsters and storage areas must be covered with tight-fitting lids, doors, or covers to discourage insects, rodents, and other types of pests.

Refuse and garbage should be removed from the site as often as necessary to prevent objectionable odors and avoid conditions that attract or harbor insects and rodents.

Outdoor storage areas must be kept clean and free of litter. Suitable cleaning equipment and supplies must be available to clean the equipment and receptacles. Refuse storage equipment and receptacles must have drains and drain plugs must be in place.

Compactors and other equipment for refuse, recyclables, and returnables must be installed to minimize the accumulation of debris. Always make sure you clean under and around these units to prevent insect and rodent harborage.

Some food establishments may provide redeeming machines for recyclables or returnables. According to the *Food*

Outside garbage and refuse station

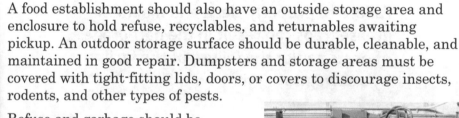

Cardboard or plastic baler

Code, a redeeming machine may be located in the packaged food storage area or consumer area of a food establishment if food, equipment, utensils and linens, and single-service and single-use articles are not subject to contamination from the machine and a public health nuisance is not created.

Pest Control

Every food establishment should have a pest control program. The targets of this program are insects and rodents that can spread disease and damage food. These pests carry disease-causing microorganisms in and on their bodies and can transfer them to food and food-contact surfaces. Pests also destroy millions of dollars of food each year by eating it or by contaminating it with urine and feces.

Integrated Pest Management (IPM)

Modern pest control operators use **integrated pest management** as the primary method to control pests in food establishments. Chemical pesticides are used only as a last resort and only in the amount needed to support the other control measures in the IPM program. Synthetic chemical pesticides are normally applied only by a Certified Pest Control Operator. If chemical pesticides are used, they must be approved for use as specified in the *Code of Federal Regulations* (40 CFR 152).

The key element of a successful pest control program is PREVENTION

However, no single measure will effectively prevent or control insects and rodents in food establishments. It takes a combination of three separate activities to keep pests in check. You must:

◆ **Prevent entry** of insects and rodents into the establishment

◆ **Eliminate food, water, and places where insects and rodents can hide**

◆ **Implement an integrated pest management (IPM) program** to control insect and rodent pests that enter the establishment.

The National Pest Management Association (NPMA) recommends a 5-step program for IPM:

1. Inspection.
2. Identification.
3. Sanitation.
4. Application of two or more pest management procedures.
5. Evaluation of effectiveness through follow-up inspections.

There are many benefits produced by using an integrated pest management program. An IPM program is more efficient and cost effective than programs that rely exclusively on chemicals to control pests. IPM is also longer lasting and safer for you, your employees, and your customers.

Insects

What insects lack in size, they more than make up for in numbers. Insects may spread diseases, contaminate food, destroy property, or be nuisances in food establishments. Insects need water, food, and a breeding place in order to survive. The best method of insect control is keeping them out of the establishment coupled with good sanitation and an integrated pest management (IPM) program when needed.

KEY TERMS

Integrated pest management (IPM)- a system that uses a combination of sanitation, mechanical, and chemical procedures to control pests.

Insects common to food establishments

Insect	Common problems	Control

Flies

Common Types:

- Houseflies
- Blowflies
- Fruit flies.

(Courtesy of Orkin)

- When a fly walks over filth, material sticks to its body and leg hairs, which can contaminate food; a fly vomits solid food to soften it before eating, spreading bacteria to food and food-contact surfaces
- Blowflies are attracted by odors in food establishments
- Fruit flies are attracted by decaying fruit.

- Eliminate the insect's food supply; store food, garbage, and other wastes in fly-tight containers; regularly clean kitchen, dining, toilet, waste storage facilities, and floor drains.
- Equip windows, doors, and loading and unloading areas with tight-fitting screens or air curtains.
- Insect electrocuting devices must be installed so dead insects and insect parts cannot fall on food and food-contact surfaces. Non-electrocuting systems, using glue traps and pheromone attractants, are allowed in areas of the establishment where electrocuting devices are not.
- Chemical insecticides may be applied by a professional pest control operator as a supplement to proper food-handling practices and a clean establishment.

Cockroaches

Common Type:

- German cockroach

(Courtesy of Orkin)

- Carry bacteria on their hairy legs and body as well as in their intestinal tract
- Commonly hide in cracks and crevices under and behind equipment and facilities.

- Maintain good housekeeping indoors and outside; eliminate hiding places by picking up unwanted materials; fill cracks and crevices in floors and walls and around equipment; doors and windows should be tight-fitting and protected by screening, air curtains, or other effective means
- Check incoming food and supplies for signs of infestation such as egg cases and live roaches; store food in containers that are insect proof and have tight-fitting lids
- Keep floors, tables, walls and equipment clean and free of food wastes
- Residual insecticides and baits can be used when a serious infestation exists.

(cont.)

Insects common to food establishments   (continued)

Insect	Common problems	Control
Moths and beetles **Common types:** ◆ Indian meal moth ◆ Saw-toothed grain beetle ◆ Flour weevil ◆ Rice weevil.	◆ These insects feed on corn, rice, wheat, flour, beans, sugar, meal, and cereals ◆ These insects create problems of wasted food and nuisance rather than disease.	◆ Inspect incoming products for signs of infestation ◆ Use FIFO system of stock rotation; store opened packages or bags of food in covered containers ◆ Clean shelves and floors frequently ◆ Keep dry food storage areas cool ◆ Residual insecticides and pheromone traps are available to control these pests.

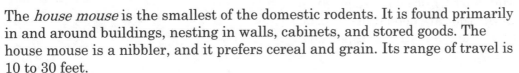

(Courtesy of Orkin)

Rodents

Rodents are known to carry microorganisms that can cause a number of human diseases including salmonellosis, plague, and murine typhus. Rodents also consume and damage large quantities of foods each year. Rats typically carry their food back to their nest rather than eat it where it is found.

Domestic rodents in the United States include the *Norway rat*, the *roof rat*, and the *house mouse*. The Norway rat is also known as the brown rat, sewer rat, and wharf rat, and is the one most commonly found in the United States.

The *Norway rat* hides in burrows in the ground and around buildings and in sewers. Norway rats will eat almost any food but prefer garbage, meat, fish, and cereal. They stay close to food and water, and their range of travel is usually no more than 100 to 150 feet.

The *roof rat* generally harbors in the upper floors of buildings but is sometimes found in sewers. Roof rats prefer vegetables, fruits, cereal, and grain for food. The range of travel for the roof rat is also about 100 to 150 feet.

Norway rat and house mouse
(Courtesy of Orkin)

The *house mouse* is the smallest of the domestic rodents. It is found primarily in and around buildings, nesting in walls, cabinets, and stored goods. The house mouse is a nibbler, and it prefers cereal and grain. Its range of travel is 10 to 30 feet.

Signs of Rodent Infestation

It is unusual to see rats or mice during the daytime since they are nocturnal. Therefore, it is necessary to look for signs of their activities. From rodent signs you can determine the type of rodent, whether it is a new or old problem, and whether there is a light or heavy infestation.

Droppings

The presence of rat or mouse feces is one of the best indications of an infestation. Fresh droppings are usually moist, soft and shiny, whereas old droppings become dry and hard. Norway rat droppings are the largest and have rounded ends. They look a lot like black jelly beans. Roof rat droppings are smaller and more regular in form. The droppings of the house mouse are very small and pointed at each end. They look something like dark grains of rice.

House mouse and Norway rat droppings

(Courtesy of Orkin)

Runways and Burrows

Rats are very cautious and repeatedly use the same paths and trails. Outdoors in grass and weeds, you may see 2- to 3-inch wide paths worn down from repeated activity.

The Norway rat prefers to burrow for nesting and harborage. Burrows are found in earth banks, along walls, and under rubbish. Rat holes are about 3 inches in diameter whereas mouse holes are only about 1 inch in diameter. If a burrow is active, it will be free of cobwebs and dust. The presence of fresh food or freshly dug earth at the entrance of the burrow also indicates an active burrow.

Rub Marks

Rats prefer to stay close to walls where they can keep their highly sensitive whiskers in contact with the wall. As a rat runs along a wall, its body rubs against the wall or baseboard. The oil and filth from the rat's body are deposited on the wall and create a black mark called a "rub mark." Mice do not leave rub marks that are detectable, except when the infestation is especially heavy.

House Mouse

(Courtesy of Orkin)

Gnawings

The incisor teeth of rats grow 4 to 6 inches a year. As a result, rats have to keep these teeth filed down in order to keep them short enough to use. Gnawings in wood are fresh if they are light colored and show well-defined teeth marks.

Tracks

Tracks may be observed along rat or mouse runs both indoors and outdoors. Look for tracks in dust in little used rooms and in mud around puddles. Rat tracks may be 1 inch long.

Miscellaneous Signs

Rodent urine stains can be seen with an ultraviolet light (black light). Rats leave a different pattern than mice. Rat and mouse hairs may be found along walls, etc. When examined under a microscope, they can be distinguished from other animal hairs.

Rodent Control

The grounds around the food establishment should be free of litter, waste, refuse, uncut weeds, and grass. Unused equipment, boxes, crates, pallets and other materials should be neatly stored to eliminate places where pests might hide.

All entrances and loading and unloading areas should be equipped with self-closing doors and door flashings to prevent rodent entry into the establishment. Metal screens with holes no larger than ¼ inch should be installed over all floor drains to prevent entry.

Traps are useful around food establishments where rodenticides are not permitted or are hazardous. Live traps can be used for collecting live rats. Check traps at least once every 24 hours. Killer or snap traps can also be used as part of a rodent control program. When using these types of traps, place them at right angles to the wall along rodent runways with the trigger side closest to the wall.

> **Effective rodent control begins with a building and grounds that will not provide a source of food, shelter, and breeding areas.**

Bait box

Bait box
(Courtesy of J.T. Easton & Co. Inc.)

Glueboard

Glueboards are shallow trays that have a very sticky surface. The mouse's feet stick to the board when it walks on it, and it is caught. Glueboards should be placed next to and running parallel with the wall.

Rodenticides are hazardous chemicals that can contaminate food and food-contact surfaces if not handled properly. Baits should be used outdoors to stop rodents at the outer boundaries of your property. Baits should be placed in a tamper proof, locked bait box that will prevent children and pets from being exposed to the toxic chemicals inside. Always make certain pesticides are stored in properly labeled containers, away from food in a secure place. Dispose of containers safely and know emergency measures for treating accidental poisoning.

The use of tracking powder pesticides is prohibited in food establishments. These types of pesticides can be dispersed throughout the establishment and directly or indirectly contaminate food, equipment, utensils, linens, and single-service/single-use articles. This contamination could adversely affect both the safety of the food and the general environment.

What Do You Think?

High School Cafeteria Reopens After Taking Measures to Correct Rodent and Roach Infestation

Health officials have given conditional approval for a local high school to reopen its lunchrooms, one week after they were shut down because of a mice and roach infestation.

Health officials closed two kitchens and a cafeteria at the school after rodent droppings were discovered in the food storage and preparation areas. Breakfast and lunch were prepared at two nearby elementary schools and shipped to the high school during the time the school cafeteria was closed.

In an effort to eliminate the pest infestation problem, school officials took a number of corrective actions. These included sanitizing affected areas, sealing holes in walls and floors, installing screens on windows, and discarding old food.

Health officials reinspected the school cafeteria and said the school could reopen the two lunchrooms provided they installed door sweeps, repaired a sink, and cleaned the storage room. These projects were completed and the cafeteria was allowed to reopen.

Summary

Back to the Story... The case presented at the beginning of the chapter illustrates how pests can pose a very serious problem for food establishments. Pests contaminate food and food-contact surfaces, and they can carry disease agents that are harmful to humans. Pests are attracted to food establishments by the food, water, and good odors they find there. In order to keep pests under control, food establishments must prevent them from entering the facility; eliminate sources of food, water, and shelter; and implement an Integrated Pest Management (IPM) program.

Customer satisfaction surveys show cleanliness is a top consideration when choosing a place to eat or shop for food. Customer satisfaction is highest in food establishments that are clean and bright and where quality food products are safely handled and displayed.

Proper construction, repair, and cleaning of floors, walls, and ceilings are important parts of an effective sanitation program. Sanitation, safety, durability, comfort, and cost are the main criteria you will use when selecting materials for floors and walls. Surfaces of floors and walls should be resistant to damage and deterioration from the water, detergents, and repeated scrubbings used to keep them clean. Walls and ceilings should be light colored to show soil and enhance the artificial lighting used in food preparation, handling, and display areas.

Handwashing stations must be properly equipped and conveniently located to enable food handlers to wash their hands as necessary throughout the workday.

Clean and suitably equipped toilet facilities must be provided for employees. These facilities must be kept clean and in good repair to prevent the spread of disease and promote good personal hygiene.

A properly designed, constructed, and installed plumbing system is very important to food sanitation. Air gaps and mechanical vacuum breakers are used to protect the municipal water supply. Consult a professional plumber or your local plumbing code for details about the plumbing requirements in your jurisdiction.

Proper storage and disposal of garbage and refuse are necessary to prevent contamination of food and equipment and avoid attracting insects, rodents, and other pests to a food establishment. Proper facilities and receptacles must be provided inside and outside the establishment to hold refuse, recyclables, and returnables that may accumulate. Refuse and garbage should be removed from food establishments frequently enough to minimize the development of objectionable odors and other conditions that attract or harbor insects, rodents, and other pests.

Pests are attracted to food establishments by the food, water and good odors they find there. In order to keep pests under control, food establishments must prevent them from entering the facility; eliminate sources of food, water, and shelter; and implement an Integrated Pest Management (IPM) program.

Quiz 7 (Multiple Choice)

Please choose the **BEST** answer to the questions.

1. The primary responsibility of food establishment managers in pest control is to assure:

    a. Good sanitation that will eliminate food, water, and harborage areas.

    b. Pesticides are applied safely.

    c. The pest control operator they use employs integrated pest management.

    d. The parking area is kept free of litter.

2. Which of the following statements about toilet facilities is **false**?

    a. Toilet facilities must be available for all employees.

    b. Employee toilet facilities must be conveniently located and accessible to employees during all hours of operation.

    c. Separate toilet facilities should be provided for men and women.

    d. Poor sanitation in toilet facilities will influence customers' opinions about cleanliness but will not promote the spread of disease.

3. The most effective device for protecting a potable water supply from backflow that would contaminate the water is a (an):

    a. Dual check valve.

    b. Hose bib.

    c. Vacuum breaker.

    d. Air gap.

4. Some signs of rodent infestation include:

    a. Rub marks along the walls.

    b. Small egg cases.

    c. Seeing mice during the day.

    d. Larva in flour and cereals.

5. What is the term used for a system that uses a combination of sanitation, mechanical, and chemical procedures to control pests?

   a. Inspection, prevention, and eradication.
   b. Individual establishment plan.
   c. Integrated pest management.
   d. Integrated poison control.

6. When an air gap is used to prevent backflow, the vertical distance between the supply pipe (faucet) and the flood rim of the sink must be at least:

   a. Two times the diameter of the supply pipe, but never less than 1/2 inch.
   b. Two times the diameter of the supply pipe, but never less than 1 inch.
   c. Three times the diameter of the supply pipe, but never less than 1.5 inches.
   d. Four times the diameter of the supply pipe, but never less than 2 inches.

7. Which of the following statements about garbage and refuse sanitation is **false**?

   a. Proper disposal of garbage and refuse are necessary to prevent contamination of food and equipment and to avoid attracting insects and rodents.
   b. A trash receptacle must be provided in each area of a food establishment where refuse is generated.
   c. The equipment and receptacles used to store refuse, garbage, etc., must be durable, clean, nonabsorbent, leak-proof, and pest proof.
   d. Trash may be stored outdoors in plastic bags provided the bags are stored at least 12 inches off the ground.

8. Backsiphonage is likely to occur if:

   a. The pressure in the potable water system drops below that of a nonpotable or contaminated water source.
   b. Contamination is forced into a potable water system through a connection that has a higher pressure than the water system.
   c. Pressure builds up in a sewer line due to blockage.
   d. The water seal in a kitchen trap is siphoned out.

9. The best way to encourage employees to wash their hands when needed is to:

   a. Provide separate restrooms for employees and customers.
   b. Provide properly equipped handwashing sinks convenient to work areas.
   c. Provide hand antiseptics instead of handwashing sinks in food preparation areas.
   d. Put up a sign in the employee locker room reminding emplooyees about the importance of proper hand washing.

10. Coving is a (an):

    a. Device used to prevent backsiphonage.

    b. Antislip flooring used to protect employees from slips and falls.

    c. Plastic material used to seal cracks and crevices under and around equipment in a food establishment.

    (d.) Curved sealed edge between the floor and wall that eliminates sharp corners to make cleaning easier.

Answers to the multiple-choice questions are provided in **Appendix A**.

Suggested Reading/References

Bennett, G, W., J. W. Owens, and R.M. Corrigan. 2003 *Truman's Scientific Guide to Pest Management Operations* (6th. ed.). Purdue University Press, West Lafayette, IN.

Code of Federal Regulations. *2010. 40 CFR 152 - Pesticide Registration and Classification Procedures.* U.S. Government Printing Office, Washington, D.C.

Food and Drug Administration and Conference for Food Protection. 2000. *Food Establishment Plan Review Guide,* Washington, D.C.

Kopanic, R. J., B.W. Sheldon, and C.G. Wright .1994. *"Cockroaches as Vectors of Salmonella: Laboratory and Field Trials," Journal of Food Protection,* 57(2), pp. 125-32.

Longrèe, K. and G. Armbruster. 1996. *Quantity Food Sanitation.* Wiley, New York, NY.

Olsen, Alan R. (1998). *"Regulatory Action Criteria for Filth and Other Extraneous Materials-Review of Flies and Foodborne Enteric Diseases," Regulatory Toxicology and Pharmacology,* 28: 199-211.

Suggested Web Sites

Actron, Inc.
  http://actroninc.com
American Society of Sanitary Engineering
  www.asse-plumbing.org
Association of Applied IPM Ecologists
  www.aaie.net/aaie/
B&G Equipment
  www.bgequip.com
Ecolab
  www.ecolab.com
EPA Office of Prevention, Pesticides and Toxic Substances
  www.epa.gov/oppts/
Hobart Corp.
  www.hobartcorp.com
Insect-O-Cutor
  www.insect-o-cutor.com
Licensed Pest Control.com
  www.licensedpestcontrol.com

National Pest Management Association(NPMA)
www.pestworld.org/

The Orkin Co.
www.orkin.com

Pest Control Industry
www.pestweb.com

PestWest, USA
www.pestwest.com

Plumbnet.com
www.plumbnet.com

The Steritech Group, Inc.
www.steritech.com

Do-It Yourself Pest Control
www.doyourownpestcontrol.com

Terminix
www.terminix.com

Virginia Tech Pesticide Programs (VTPP)
www.vtpp.ext.vt.edu

NSF International
www.nsf.org

Underwriters Laboratories, Inc.
www.ul.com

National Restaurant Association
www.restaurant.org

The Food Marketing Institute
www.fmi.org

Gateway to Federal Food Safety Information
www.foodsafety.gov

U.S. Department of Agriculture (USDA)
www.usda.gov

U.S. Food and Drug Administration (FDA)
www.fda.gov

Notes

Food Safety Management Programs and the HACCP System

Key Terms

- Coding
- Critical Control Point (CCP)
- Critical Limit (CL)
- Food recall
- Good Retail Practices (GRPs)
- Hazard Analysis Critical Control Point (HACCP)
- Risk
- Standard Operating Procedures (SOPs)

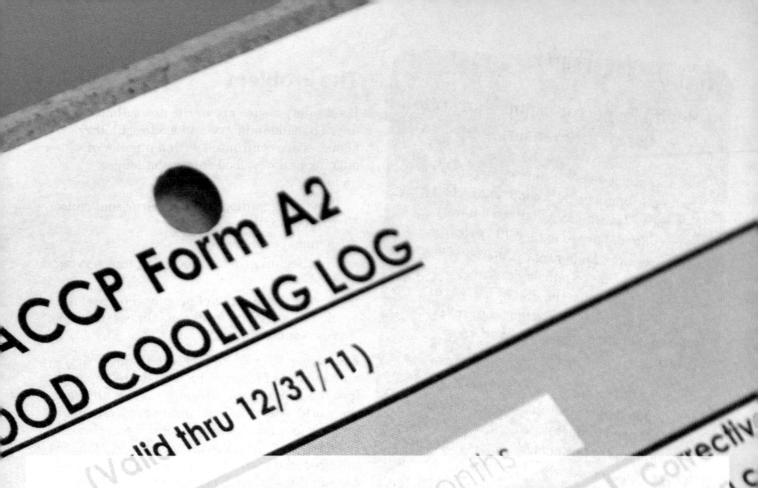

In previous chapters of this book, you have learned about the factors that can lead to both foodborne illness and food spoilage. You have also learned some of the key preventive measures and controls that can be used to minimize contamination of foods throughout the flow of food. **The focus of this chapter is to understand how to build effective food safety management programs**, including HACCP plan development, within your food establishment.

Learning Points:

◆ Recognize the usefulness of effective food safety management using the HACCP system.

◆ Recognize the types of PHF (TCS) products that benefit from a HACCP system to assure food safety.

◆ Identify the steps involved in implementing a HACCP plan.

◆ List and assess hazards and risk factors related to each product analyzed.

◆ Identify points, called critical control points, in the flow of food to be monitored.

◆ State measures used to correct potential problems.

◆ Identify data required to provide documentation for review of the HACCP system.

◆ Apply the HACCP system to analyze and protect PHF(TCS) items from contamination during processing, preparation, and service.

◆ Describe the process used to verify the HACCP system is working effectively.

◆ Identify and review procedures for crisis management situations due to loss of potable water and for foodborne illness complaints.

The Problem

Food safety issues are in the news almost every day. The food industry and food regulatory agencies are confronted with a number of ongoing and new food safety challenges, including:

- Emerging pathogens or hazards that cause foodborne illness
- A global food supply
- New techniques for processing and serving food
- A growing number of people who are classified as at-risk to foodborne illness (immune-compromised population).

Every year there is an increase in the variety and amount of food products that are imported into the United States from countries all over the world. Also, methods used to process and prepare foods domestically continue to change. In addition, governmental agencies at all levels have reduced staff and services due to increased costs and fewer local, state, and federal funds. Because of these factors, increased responsibility for food safety has shifted to food establishment managers and employees.

The Solution

Chapter 3 identifies the major risk factors related to foodborne illness. The most important factors are:

- Improper holding temperatures
- Poor personal hygiene
- Contaminated equipment
- Inadequate cooking
- Food obtained from an unsafe source.

Food safety management programs involve:

- Temperature and time control
- Good personal hygiene practices and procedures
- Preventing cross contamination
- Effective cleaning and sanitizing
- Obtaining food from an approved source.

Food safety management programs should protect foods from these risk factors known to lead to foodborne illness. To ensure food safety, we need to focus on the following keys to foodborne illness prevention:

- Temperature and time control.
- Good personal hygiene practices and procedures.
- Preventing cross contamination.
- Effective cleaning and sanitizing
- Obtaining food from an approved source

Food safety management programs in food establishments must be customized to fit company policies, the size of the establishment, the products sold and the number and types of customers served. To operate a food safety management program, you must understand the hazards that cause foodborne illness, and how these hazards can be prevented. Focus the program on elimination or control of the risk factors known to contribute to foodborne illness. The food safety procedures in your management system must be easy for your employees to understand and implement. Management in your operation will be up to you.

The Hazard Analysis Critical Control Point (HACCP) System

In the food industry, there are a variety of programs that can be used to manage food safety. The **Hazard Analysis Critical Control Point (HACCP)** system is a prevention-based program used by many segments of the food industry to assure food safety. The HACCP system can control hazards, such as disease-causing microorganisms, chemical substances, and physical objects that are common causes of foodborne illness.

In 1959, the National Aeronautic and Space Agency (NASA) approached the Pillsbury Company to produce food that was safe for astronauts to eat in zero gravity conditions in the space capsules. The project goal was to develop a system that assured nearly 100% freedom from contamination by microbial, chemical, or physical hazards. The process worked so well it was quickly introduced into the food industry as a key food safety and protection tool.

Product Description and Development of the Food Flow Diagram

Before an effective food safety management program can be developed, you need to know important information about the product you are making and the process that is used for food preparation. We often say that food safety management programs are "product and process specific."

To better understand the product that you are making, it is useful to describe the product by listing all of the ingredients that are incorporated into the food – or the product ingredient list. We will use this list of ingredients to identify which ingredients are PHF (TCS). This helps us focus best food-handling practices and controls on high-risk products.

An example of a product description for meatloaf shown as an ingredient list is presented below.

Ingredients	Amount	Servings		
		25	50	100
Ground beef	Lbs.	6	12	24
Onion, diced	Cups	2	4	8
Green pepper, diced	Cups	1	2	4
Celery, diced	Cups	1	2	4
Bread crumbs	Cups	3	6	12
Pasteurized eggs, frozen	Cups	1	2	4
Catsup	Cups	1	2	4
Homogenized milk	Cups	½	1	2
Pepper	Cups	½	1	2
Salt	Cups	½	1	2

Ingredient list for meatloaf

HACCP is not required for food establishments except in a limited number of states and jurisdictions, yet many food establishments have voluntarily implemented HACCP principles to assure the safety and wholesomeness of the products they produce and sell.

The next step is to describe the process used to prepare a food by developing a food flow diagram. We learned about food flow diagrams in Chapter 4 of this book. The food flow diagram identifies all of the storage and preparation steps that are involved in making and serving the food. We will use the food flow diagram to identify which steps in the process are most important to control the food safety hazards that were described in Chapter 2 of this book.

An example of a food flow diagram for preparation of meatloaf appears below.

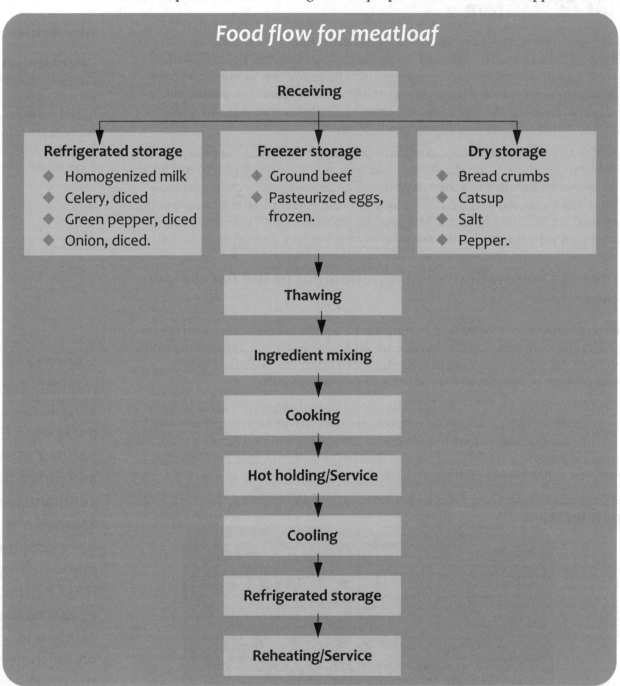

Food flow for meatloaf

Receiving

Refrigerated storage
◆ Homogenized milk
◆ Celery, diced
◆ Green pepper, diced
◆ Onion, diced.

Freezer storage
◆ Ground beef
◆ Pasteurized eggs, frozen.

Dry storage
◆ Bread crumbs
◆ Catsup
◆ Salt
◆ Pepper.

Thawing

Ingredient mixing

Cooking

Hot holding/Service

Cooling

Refrigerated storage

Reheating/Service

After the product and process have been described, the more formal process of developing a HACCP plan can be done. As we have just described, the initial step of developing a HACCP plan is to understand the food formulation and the flow of food that will be used to produce each food item. Having an understanding of the food ingredients and the process used to make the food is the foundation for developing a HACCP plan. HACCP plan development is composed of 7 steps which are described in the next several pages of this chapter.

In addition to the HACCP system, there are many other programs and practices that need to be done to protect the safety of foods. These practices complement the HACCP systems and assure that the HACCP plan will work effectively. These include:

◆ Standard Operation Procedures (SOPs) that assure uniform food safety compliance
◆ Use of approved products
◆ Facility design that meets code requirements
◆ Employees educated and supervised in good personal hygiene practices
◆ An effective cleaning and sanitation program
◆ Proper equipment and maintenance programs
◆ A commitment from management to facilitate the HACCP system.

In a HACCP food safety system, the focus is on food and how it is handled during storage, preparation, display, and service. A sanitary environment is important for safe food production; however, food can still be contaminated and microbes can grow if:

◆ Proper food-handling techniques are not used
◆ Good personal hygiene is not practiced
◆ Cross contamination controls are not in place
◆ Food temperatures are not controlled.

It is important to understand that the HACCP approach should be used to help food managers identify and control potential problems <u>before</u> they happen. It is only effective when tailored to the specific needs of the food establishment. HACCP should not be viewed as a "one size fits all" program. Each HACCP plan is specific to the product being produced and the process that is being used to make the food. Whether used in restaurants, retail food establishments, institutions, health-care facilities, or other foodservice operations, the primary goal is always the same – production of safe and wholesome food.

HACCP is the preferred approach to food safety because it provides the most effective and efficient way to assure food products are safe. By using the HACCP system, food managers can identify the foods and processes that are most likely to cause foodborne illnesses. When a potential problem is identified, the food establishment can initiate procedures to reduce or eliminate the risk of foodborne illness and monitor actions to make sure the procedures are being followed.

The HACCP system for food operations normally uses control of time, temperature, and specific factors that are known to contribute to foodborne disease outbreaks. Records produced in conjunction with the HACCP system provide a comprehensive source of information about the events that occurred during all stages of food production.

Food managers and supervisors must learn how to effectively develop, implement, and maintain a HACCP system. There are several publications available, including an excellent guide in the *Food Code* that can help you learn more about the HACCP system. Consult the suggested readings at the end of this chapter for a partial list of these publications.

In a HACCP food safety system, the focus is on food and how it is handled during storage, preparation, display, and service.

Good personal hygiene is an important prerequisite for a successful HACCP program.

The Seven Principles in a HACCP System

The basic process in the development of a HACCP plan consists of seven principles as listed below. Remember that the HACCP plan is developed after the food product and food process have been clearly identified. Also remember that other programs, like SOP's, will be used to complement the HACCP approach. While each principle is unique, they all work together to form the basic structure of an effective food safety management program. A more detailed description of each principle will be presented in the remainder of this chapter to help you understand better how the system works.

The HACCP system is an important part of a food safety management program.

Seven principles in a HACCP system

1. Hazard analysis.

2. Identify the critical control points (CCPs) in food preparation.

3. Establish Critical Limits (thresholds) that must be met at each identified CCP.

4. Establish procedures to monitor CCPs.

5. Establish the corrective action to be taken when monitoring indicates a critical limit has been exceeded.

6. Establish procedures to verify the HACCP system is working.

7. Establish effective record keeping that will document the HACCP system.

Principle 1—Hazard Analysis

The first principle in a HACCP system is the hazard analysis. Hazard analysis starts with a thorough review of your menu or product list to identify all the PHF (TCS) products you serve. As you learned in *Chapter 2, Hazards to Food Safety*, PHF (TCS) require temperature and time control to limit pathogenic microorganism growth and toxin formation. PHF (TCS) include:

◆ Foods of animal origin that are raw or heat-treated
◆ Foods of plant origin that are heat-treated or consist of raw seed sprouts
◆ Cut melons
◆ Garlic and oil mixtures that are not modified in a way to inhibit the growth of microorganisms
◆ Cut tomatoes including sliced, diced, chopped, and pureed tomatoes
◆ Cut leafy greens (lettuce, spinach, and salad mixes).

All of these foods are commonly found as ingredients or finished foods in food establishments.

Remember from Chapter 2 that foodborne hazards fall into three basic categories including biological, chemical, and physical hazards. We learned that the biological hazards are the most important hazards to control within food establishments. The main goal of the hazard analysis step is to connect these potential hazards with the PHF (TCS) items that you produce. Go back to chapter 2 and see if you can identify possible hazards that could be associated with the ingredients of the meatloaf. What additional hazards do you think could occur during the preparation of the meatloaf?

During the hazard analysis step, it is important to estimate risk. **Risk** is the probability that a condition or conditions will lead to a problem. Risk is an estimate of the likely occurrence of a hazard. The problem in this case is foodborne illness. Some of the factors that influence risk are the:

KEY TERM

Risk - the probability that a condition or conditions will lead to a problem.

◆ Type of customers served
◆ Ingredients used in the food formulation
◆ Types of foods on the menu
◆ Nature of the foodborne hazard
◆ Extent of food handling
◆ Information from past foodborne outbreaks
◆ Size and type of food production operations
◆ Extent of employee training.

Each type of operation and each food establishment poses different levels of risk to the customer. Focus on the ingredients and foods that have the highest risk first. Once your HACCP system has been developed, implemented, and evaluated for these foods, move on to the next most hazardous foods.

Hazard identification and risk estimation provide a logical basis for determining which hazards are significant and must be addressed in the HACCP plan. The severity of a hazard is defined by the degree of seriousness of the consequences should it become a reality. Hazards that involve low risk are less important and may not need to be addressed using a HACCP plan.

Personal experience, facts generated by foodborne illness investigations, and information from scientific articles can be useful when estimating the approximate risk of a hazard. When estimating risk, it is important to separate food safety concerns from food quality issues. For instance, the fact that perishable foods spoil quickly when stored in the food temperature danger zone is a quality issue, not a food safety issue. However, foodborne disease investigations show that allowing PHF (TCS) items to remain in the temperature danger zone too long is a common contributor to foodborne illness.

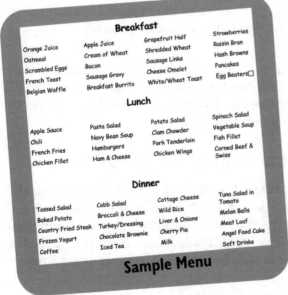

Sample Menu

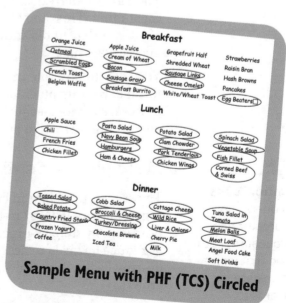

Sample Menu with PHF (TCS) Circled

Identify the PHF (TCS) foods and ingredients you serve.

The last phase of the hazard analysis step involves establishing preventive measures to control the hazards. We can control hazards by destroying them (i.e., cooking), preventing them from growing (i.e., hot-and cold-holding), and minimizing contamination (i.e., foods from approved suppliers, good hygiene practices, cross contamination control). After the hazards have been identified, you must consider what preventive measures can be employed to best control each identified hazard.

We have already learned that the key preventive measures used in food establishments include:

◆ Controlling the temperature and time of the food
◆ Cross contamination control
◆ Good personal hygiene practices
◆ Obtaining food from an approved source
◆ Having an effective food safety management program in place.

Your HACCP system should employ preventive measures that can be easily monitored. Since food temperature and time can be easily monitored, they are the preventive measures used most often in a HACCP system for food establishments.

In the meatloaf example at the beginning of this chapter, consider all of the hazards that you have learned about in previous chapters could be associated with this food product. For example, we have learned that biological hazards such as *Salmonella* and *E. coli* bacteria are often associated with raw meat. Ground meat can also introduce physical hazards, like bone fragments. The refrigerated diced vegetables can introduce bacteria like *Salmonella* and chemical hazards such as pesticide residues. Dried spices can introduce sporeforming bacteria, like *Clostridium perfringens* and *Bacillus cereus*.

Then as the process of making meatloaf continues after the cooking process, temperature control will be important to prevent the growth of bacteria that survived the cooking process (i.e., sporeforming bacteria) and for those that may be introduced during handling (i.e., *Staphylococcus aureus* from contact with food employees).

The hazard analysis step in a HACCP system is difficult and can be time consuming. This step of the process often requires the experience and expertise of a team that understands survival, growth and inactivation of microorganisms.

Your day-to-day role in the HACCP system may not involve hazard analysis. Rather, your involvement will likely focus more on other steps in the HACCP process, such as measuring temperature and time during the food product flow and taking corrective action when a problem arises.

Principle 2—Identify Critical Control Points (CCPs)

The second principle in creating a HACCP system is to identify the **critical control points (CCPs)** in food production. A CCP is "an operation (practice, preparation step, or procedure) in the flow of food that will prevent, eliminate, or reduce hazards to acceptable levels." A CCP provides a control step that will destroy bacteria or prevent or slow down the rate of disease-causing bacterial growth.

Examples of CCPs include:

◆ Cooking, reheating, and hot-holding

◆ Chilling, chilled storage, and chilled display

◆ Receiving, thawing, mixing ingredients, and other food-handling stages

◆ Product formulation of manufactured goods (i.e., pH of a food at or below 4.6 or A_w at 0.85 or below)

◆ Purchasing seafood, MAP foods, and ready-to-eat foods (where further processing would not prevent a hazard) from approved sources.

Hot-holding is a common CCP.

The most commonly used CCPs in food establishments are cooking, cooling, reheating, hot-holding, and cold-holding. Cooking and reheating to proper temperatures will destroy bacteria, whereas proper cooling, hot-holding, and cold-holding will prevent or slow down the rate of bacterial growth.

CCPs

CCPs involve:

◆ Time

◆ Temperature

◆ Acidity

◆ Water activity (A_w)

◆ Purchasing and receiving procedures related to:

  ◆ Seafood

  ◆ Modified atmosphere packaged foods

  ◆ Ready-to-eat foods where a later processing step in the food flow would not prevent a hazard

◆ Thawing of ready-to-eat foods where a later processing step in the food flow would not prevent a hazard.

> There must be at least one critical control point in the production process to qualify it as a HACCP food safety system, and the critical control point must be monitored and controlled to assure the safety of the food.

Cross contamination control, employee hygiene, and environmental hygiene are more difficult to measure, monitor, and document. Therefore, many food establishment operators prefer to think of them as **"standard operating procedures" (SOPs)**, also called "best practices" or "internal policies" rather than CCPs. SOPs are written procedures on how to perform a job.

SOPs

SOPs Include:

◆ Good employee hygiene practices (i.e., hand washing)

◆ Cross contamination control (i.e., keeping raw products separate from cooked and ready-to-eat foods)

◆ Environmental hygiene practices (i.e., effective cleaning and sanitizing of equipment and utensils).

Identification of critical control points begins with a review of the recipe for the PHF (TCS) based ingredients and the development of a food flow chart for the recipe. The flow chart tracks the steps in the food flow from receiving to serving or sale. The specific path a food follows will be slightly different for

each product. However, some of the more common elements in the flow of food include:

♦ Purchase of products and ingredients from sources inspected and approved by regulatory agencies
♦ Receiving products and ingredients
♦ Storage of products and ingredients
♦ Preparation steps which may involve thawing, cooking, and other processing activities
♦ Holding or display of food
♦ Service of food
♦ Cooling food
♦ Storing cooled food
♦ Reheating food for service.

Controlling the temperature of food products throughout the flow of food is the most commonly used technique for assuring food safety for food establishments. However, time is an important factor that can also be used as a public health control measure. As you learned in Chapter 2, it commonly takes 4 hours or more in the temperature danger zone for bacteria to multiply to levels where they will cause foodborne illness.

On the next page, the same food flow diagram is presented for the production and service of meatloaf, except that we have included examples of 5 different SOPs (labeled as SOP-1, SOP-2, SOP-3, SOP-4, and SOP-5) and 5 different CCPs (labeled as CCP-1, CCP-2, CCP-3, CCP-4, and CCP-5). This is an example of how a food safety management program could be developed for meatloaf production using HACCP principles and complementary SOPs. Notice how each CCP is dedicated to control of temperature <u>and</u> time.

Principle 3—Establish the Critical Limits Which Must Be Met at Each Critical Control Point

KEY TERM

Critical Limits (CLs) - the maximum or minimum value to which a hazard must be controlled at a critical control point to minimize the risk that the identified food safety hazard may occur.

This principle involves setting **critical limits (CLs)** to make sure each CCP effectively controls a biological, chemical, or physical hazard. CLs should be thought of as the upper and lower boundaries of food safety. The CL should be as specific as possible, such as "heat ground meat to an internal temperature of 155°F (68°C) or more for at least 15 seconds." A well-defined CL makes it easier to determine when the limit has not been met.

Each CCP has one or more CLs to monitor to assure hazards are:

♦ Prevented
♦ Eliminated
♦ Reduced to acceptable levels.

Each limit relates to a process that will keep food in a range of safety by controlling:

♦ Temperature
♦ Time
♦ The ability of the food to support the growth of infectious and toxin-producing microorganisms.

Criteria most frequently used for critical limits

Critical Limit	Boundaries of Food Safety
Temperature	◆ Keep PHF (TCS) at or below 41°F (5°C) or at or above 135°F (57°C). Maintain specific cooking, cooling, reheating, and hot-holding temperatures.
Time	◆ Limit the amount of time food is in the temperature danger zone during preparation and service processes to 4 hours or less.
Water activity	◆ Foods with a water activity (A_w) of .85 or less do not support growth of disease-causing bacteria.
pH (acidity level)	◆ Disease-causing bacteria do not grow well in foods that have a pH of 4.6 or below.

Food flow diagram for meatloaf showing CCPs, SOPs and CLs for each CCP

SOP – 1	**Receiving** ◆ Inspect all ingredients upon receiving to ensure that they are acceptable. ⬇
SOP – 2	**Refrigerated Storage, Freezer storage, Dry storage** ◆ Store all ingredients under the correct temperature conditions. ⬇
SOP – 3	**Thawing** ◆ Thaw frozen ground beef and liquid eggs under refrigeration at 41°F (5°C) or below. ⬇
SOP – 4	**Ingredient mixing** ◆ Clean and rinse vegetables under cool running water ◆ Combine all ingredients into mixer and blend for 5 minutes. ⬇
CCP – 1	**Cooking** ◆ Cook to an internal temperature of at least 155°F (68°C) for 15 seconds hold time. ⬇
CCP – 2	**Hot holding/Service** ◆ Cover and hold at 135°F (57°C) or higher. ⬇
CCP-3	**Cooling** ◆ Cool from 135°F (57°C) to 70°F (21°C) within 2 hours and from 135°F (57°C) to 41°F (5°C) within 6 hours. ⬇
CCP – 4 SOP – 5	**Refrigerated storage** ◆ Store unused meatloaf in 2" deep pans and refrigerate at 41°F (5°C) or below ◆ Cover, label, and date containers. ⬇
CCP – 5	**Reheat/Service** ◆ Reheat to an internal temperature of 165°F (74°C) or above, within 2 hours.

If we look back to the food flow diagram presented for meatloaf, 5 CCPs were identified. Each of these CCPs has a CL that involves a measurement of temperature and time. We obtained the CLs from what we learned in chapters 3 and 4 of this book. To summarize, we have 5 CCPs, and a set of CLs for each CCP, that would need to be monitored that include:

CCP	CL's
Cooking	Temperature: 155°F (68°C) or above Time: 15 seconds
Hot-holding	Temperature: 135°F (57°C) or above Time: Throughout hot-holding time period
Cooling	Temperature: 135°F (57°C) to 70°F (21°C) Time: 2 hours or less, AND; Temperature: 135°F (57°C) to 41°F (5°C) Time: 6 hours or less
Refrigerated storage	Temperature: 41°F (5°C) or below Time: Throughout cold-holding time period
Reheat	Temperature: 165°F (74°C) Time: Within 2 hours.

Principle 4—Establish Procedures to Monitor CCPs

Staff members must be given the responsibility for monitoring CCPs. This involves making observations and measurements to determine whether a CCP is under control. Monitoring will show when a CCP has exceeded its CL.

Time and temperature are the critical limits most commonly monitored to assure a critical control point is under control for food establishments. The pH and A_w of food may also be used in some cases. If a product or process does not meet critical limits, immediate corrective action is required before a problem occurs.

Monitoring must be "doable." Frequent monitoring catches problems early and provides more options for correction (i.e., reheat instead of discard food that is not maintained at current hot-holding temperatures). If a CL cannot be monitored continuously, set up specific monitoring intervals that can accurately indicate hazard control. If early monitoring indicates the process is very consistent, it is not necessary to monitor as often.

Observations of cross contamination control, employee hygiene compliance, and product formulation control should also be incorporated into the monitoring system. A record of actions, times, temperatures, and any departure from CLs provides a history for that item. Once the flow is established, measurements can be recorded on a flow chart. Have employees place their initials by the information they record.

Establish guidelines for measuring food temperatures

Teach food employees responsible for monitoring CCPs how to accurately measure CCPs and record the information in data records. Explain to employees about the harm that can occur if data is not collected as required.

Monitoring is a critical part of a HACCP system. It provides written documentation that can be used to verify the HACCP system is working properly. An operation that identifies critical control points and establishes CLs without having a monitoring system in place has not actually implemented a HACCP system. CLs without proper monitoring are meaningless.

Teach employees the importance of following guidelines

In the HACCP example for meatloaf, we learned that there were 5 CCPs that needed to be monitored. The 5 CCPs were cooking, hot-holding, cooling, cold storage, and reheating. For each of these steps, it will be important that you establish a protocol for accurately measuring temperature throughout the flow using a properly calibrated temperature-measuring device. Be sure to record temperature measurements on a temperature log.

Principle 5—Establish the Corrective Action to Be Taken When Monitoring Shows a Critical Limit Has Been Exceeded

If you detect a CL was exceeded during the production of a HACCP monitored food, correct the problem immediately. The flow of food should not continue until CLs for all CCPs have been met.

First, determine what went wrong. Next, choose and apply the appropriate corrective action. For example, if the temperature of the meatloaf during hot-holding was not maintained at 135°F (57°C) or higher, check to be sure that the hot-holding equipment is working properly. At the same time, put the meatloaf in the oven to reheat it to 165°F (74°C). The meatloaf should be discarded if you suspect it has been in the temperature danger zone for more than 4 hours.

Taking immediate corrective action is vital to the effectiveness of your food safety system.

> **Take action if a critical limit has been exceeded.**

Principle 6—Establish Procedures to Verify the HACCP System Is Working

The sixth principle in the HACCP system is to verify your system is working properly. First, verify the CLs you have established for your CCPs will prevent, eliminate, or reduce hazards to acceptable levels. Second, verify the overall HACCP plan is functioning effectively. Are good records in place? Is corrective action being taken when necessary? The HACCP system should be reviewed and, if necessary, modified to accommodate changes in:

> **Review HACCP procedures periodically.**

◆ Clientele (i.e., increase in high-risk populations that you serve)

◆ The items on the menu or ingredient list [especially PHF (TCS)] that change

◆ The processes used in the flow of food to prepare foods that have a HACCP plan.

Your management team should review and evaluate the establishment's HACCP program at least once a year, or more often if necessary.

Principle 7—Establish an Effective Record-Keeping System That Documents the HACCP System

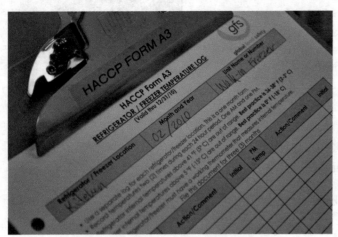

Keep sufficient records to prove your HACCP system is working effectively.

An effective HACCP system requires the development and maintenance of a written HACCP plan. The plan should provide information about the hazards associated with individual food items or group of food items covered by the system. Clearly identify each CCP and the critical limits that have been set for each CCP. The procedures for monitoring critical control points and recording maintenance must also be contained in the establishment's HACCP plan.

The amount of record keeping required in a HACCP plan will vary from one food establishment to another depending on the type of food processing used. The details of your HACCP plan will be determined by the complexity of your food production operation. Keep sufficient records to prove your system is working effectively, but keep it as simple as possible.

Changing a procedure at a CCP but not recording the change on your flow chart almost guarantees similar problems will be repeated. Record keeping is vital to the overall effectiveness of your HACCP system.

A clipboard, work sheet, thermometer, clock, and any other equipment needed to monitor and record these limits must be readily available to the food production staff. The method you use to record the information is not especially important, as long as it is easy for the staff to use and provides quick access to the information contained in the record log.

Temperature Log			
CCP – 1: Meatloaf must be cooked to an internal temperature of at least 155°F (68°C) for at least 15 seconds.			
Date	Time of Day	Cooking Temperature	Temperature Checked by
2/20/10	8:41 am	172°F	ALD
2/20/10	9:13 am	161°F	ALD
2/20/10	3:34 pm	159°F	RLT
2/20/10	3:50 pm	176°F	RLT
2/20/10	4:10 pm	170°F	RLT

An example of a temperature log for monitoring cooking temperatures of meatloaf

Roles and Responsibilities Under HACCP

The role of health department personnel and other regulators is to promote the use of HACCP and its principles by the food industry. Regulatory personnel can review your HACCP documents periodically to ensure CCPs are properly identified, CLs are properly set, required monitoring is being performed, corrective action is being taken, and the HACCP plan is being revised when necessary.

The job of the food establishment managers and supervisors is to develop, implement, and maintain the HACCP system. Continuously use and improve your HACCP system to achieve safe food management.

Complementary Food Safety Management Programs

HACCP is the most recognized food safety management system in the world, but it is not a stand-alone system. HACCP requires that other effective food safety and food quality systems are working. We call these programs prerequisite programs. In the food manufacturing industry, Good Manufacturing Practices (GMPs) are prerequisite programs that focus on the minimum sanitary and processing requirements necessary to assure the production of safe and wholesome food. GMPs include provisions for food handling related to personnel, building and facilities, equipment and utensils, and production and process controls. GMPs help to control contamination by implementing good personal hygiene and effective cleaning and sanitizing procedures. Provisions for GMPs in food manufacturing can be found in Section 21 of the *Code of Federal Regulations*, Part 110.

A similar concept has been undertaken for foods that are produced on the farm as well as for food prepared in food establishments. Good Agricultural Practices (GAPs) are used to reduce the risk of contamination of fruit and vegetable products on the farm. Producers that use GAPs focus on farm management programs, such as personal hygiene, irrigation water, manure, and cleaning procedures for freshly harvested produce. GAPs are important for food establishments too. Assuring your produce vendors adopt GAP programs will be very important for reducing the risk of ready-to-eat produce used and sold in your establishment.

Similarly, the retail food industry embraces this concept with **Good Retail Practices (GRPs)**. GRPs are preventive measures that include practices and procedures which effectively control the introduction of pathogens, chemicals, and physical objects into food. In food establishments, GRPs are the key prerequisites to instituting a HACCP plan or risk control plan. Prerequisite programs may include such things as:

◆ Vendor certification programs
◆ Training programs
◆ Allergen management
◆ Buyer specifications
◆ Recipe/process instructions
◆ First-In-First-Out (FIFO) procedures
◆ Other Standard Operating Procedures (SOPs).

KEY TERM

Good Retail Practices (GRPs) - preventive measures that include practices and procedures which effectively control the introduction of pathogens, chemicals, and physical objects into food.

Basic prerequisite programs should be in place to:

◆ Protect products from contamination by biological, chemical, and physical food safety hazards
◆ Control bacterial growth that can result from temperature abuse
◆ Maintain equipment.

Sell-by date

Coding and Product Identification

The product code for manufactured goods and for goods that are prepared and packaged in food establishments is an important way to identify

Use-by date

the product and when the product was produced. In a food establishment, product coding is customized to the operation. But, the product code needs to provide information about the product and when it was produced.

One of the more common uses of product coding in retail is for short shelf-life foods that are refrigerated. Product coding (in the form of "Best By," "Sell By," and "Best if Used By") can be used to assure foods are displayed and used on a FIFO basis and that products with an expired shelf life are discarded. See Chapter 9 for more details on food labeling requirements.

Sell-by date from a food product prepared in a deli operation

Food Recalls

On occasion, contaminated products may be received from manufacturers, slaughterhouses, or farms that can present a risk to your customers. In these situations, the company that supplied the contaminated product may put a **food recall** in place. A food recall is a process that is implemented to remove a defective product that presents a risk of injury or gross deception. An important part of a recall procedure is to communicate information about the recalled products to staff that work in the facility so that the product is not used (i.e., restaurant), and to customers that may have purchased the product and taken it home (supermarket). Another important part of a food safety management system is to develop an effective food recall procedure.

There are three types of food recalls depending on the severity of the risk a product may present. FDA or USDA identifies recalls as Class I, Class II, or Class III. Food establishments should have a policy in place to handle food recalls. Recalled foods should be promptly removed from supermarket shelves and from restaurant storage areas and cannot be sold or used in foods.

Types of food recalls

Class I - foods that may cause serious adverse health consequences

Class II - foods that would result in a temporary or reversible health problem

Class III - foods that are not likely to cause danger to health.

FDA or USDA can request a food recall if the agency has determined a potential hazard exists from a contaminated food. In the majority of cases, food processors voluntarily remove products from the marketplace to keep consumers safe.

Food Safety Related Crisis Management Situations

In the fast-paced world of retail food production, a crisis can occur at any time. Besides natural disasters such as fire, flood, storms, or earthquakes, food establishments must deal with interference in normal operations that can impair production. Examples of crises that can have a significant impact on food operations are power outages, interruption of water and sewer service, an outbreak of foodborne illness, media investigations, and unexpected loss of personnel. How do you handle difficult situations? Develop a plan to handle the crisis before it happens. Use a problem-solving approach to develop a response plan and test it to make sure it works.

Evaluate the extent of the problem. Set priorities according to the resources that exist. Remain calm. Follow established rules and regulations as set up in codes. Identify any outside resources that can help solve the problem. Keep a record of actions and communications in case it should be needed. In all situations, be honest.

Water Supply Emergency Procedures

On occasion, a water utility will issue a boil water order when the bacteriological quality of the water it provides does not meet the requirements of the Safe Drinking Water Act. This commonly happens when a water main breaks or when the source of the drinking water is contaminated by runoff and other sources of pollution.

To continue operating under "boil water advisories/notices" or "interrupted water service" from all water supplies, food establishments must secure and use safe drinking water from an approved source. This can be bottled water or water hauled to the site in tank trucks. Bottled drinking water used or sold in a food establishment must be obtained from approved sources in accordance with the Code of Federal Regulations (21 CFR 129—*Processing and Bottling of Bottled Drinking Water*).

Have alternative sources of water available during a water emergency.

In emergencies, or as a temporary measure, water from contaminated or suspect sources can be disinfected by either boiling or chlorination. The Environmental Protection Agency (EPA) says boiling is the surest method to make water safe to drink and kill disease-causing microorganisms. Water can be made safe by bringing it to a full boil for at least one minute. Once the water has been boiled it can be cooled and aerated by pouring it from one clean and sanitary container to another. This will reduce the flat taste caused by boiling.

To disinfect water using household bleach, add 8 drops of liquid chlorine bleach to 1 gallon of water and mix. Check the label on the bleach container to make sure the active ingredient, sodium hypochlorite, is available at 5.25%. The important things to remember when chlorinating water are:

◆ Mix the treated water thoroughly and allow it to stand for 30 minutes before using it for drinking or cooking.

◆ If 8 drops of bleach do not give the water a slight taste of chlorine, add 8 more drops until a slight taste is present. The taste of chlorine will be evidence the water is safe to drink.

More detailed information from the EPA about emergency disinfection of drinking water can be found at www.epa.gov/safewater/faq/emerg.html.

Alternative procedures to minimize water usage

◆ Commercially packaged ice may be substituted for ice made on-site
◆ Single-service items or disposable utensils may be substituted for reusable dishes and utensils
◆ Use prepared foods from approved sources in place of complex preparation onsite
◆ Restrict menu choices or hours of operation
◆ Portable toilets may be utilized for sanitary purposes.

In emergencies, or as a temporary measure, water from contaminated or suspect sources can be disinfected by either chlorination or boiling.

After the water emergency is officially lifted or water service resumes, these precautionary measures must be followed:

◆ Flush the building water lines and clean faucet screens and waterline strainers on mechanical dishwashing machines and similar equipment
◆ Flush and sanitize all water-using fixtures and appliances such as ice machines, beverage dispensers, hot water heaters, etc.
◆ Clean and sanitize all fixtures, sinks, and equipment connected to waterlines.

There must be adequate water pressure and the bacteriological quality and safety of the water supply must be verified before a food establishment can resume operations. If a food establishment is on a public water supply, the water provider will be responsible for having the bacteriological quality of the water tested. If the water in a food establishment is supplied by a private well, then it is the establishment's responsibility to have the water tested. Either way you must remember that the safety of drinking water cannot be judged by color, odor, or taste. Don't hesitate to contact your local regulatory agency if you have any questions regarding appropriate operations at your establishment during and after a water supply emergency.

Foodborne Illness Incident or Outbreak

If dealing with a suspected foodborne illness problem, cooperate with the customers and seek help from your local food regulatory agency. Work with the local regulatory agency to determine what may have caused the customer's illness. Establish the time the suspect food was purchased and eaten. If a sample of the product is still available, preserve it for laboratory analysis. Ask the customer if medical attention has been sought and to save samples of any product that remains. Initiate your crisis management team, which should include members from media affairs, food safety, loss prevention, and legal counsel. If you do not have these resources available on your staff, you can hire them as consultants to work with your establishment.

Action plan when foodborne illness is suspected

- Remove all affected food
- Isolate it in a separate area
- Label it so it does not become mixed with other foods
- Do not sell any more of the suspect product
- Listen to everything the customer says and look for clues that might explain the problem
- Do not try to diagnose the problem or belittle the customer's complaint
- Keep a record of the interview
- Assure the customer you will get back to him or her after investigating the problem
- Make sure you do contact the customer again after facts are gathered
- Do not admit your establishment is at fault—you have no proof of that until facts are checked.

If the reported item was prepared in the food establishment, remove it from sale immediately. If two or more persons report the same problem, consult outside resources (such as the local health department) on how to proceed. When preparing food samples for testing, it is wise to keep a sample in your freezer for cross validation of laboratory results. Public health officials, insurance agents, and, if needed, an attorney can help you manage the problem.

If there is a formal investigation, remove suspect items from use. Health inspectors may want to conduct an inspection at the establishment. Employees may be required to have medical examinations. Closing a facility for a period of time for cleaning may be needed. Keep in mind whatever must be done should be accomplished as quickly as possible. Cooperation to protect public health is the end goal. In the final analysis, it does not matter where the error occurred. It must be corrected and prevented from happening again.

Whatever the crisis may be, use a calm, systematic approach and evaluate actions as much as possible. Crisis management is not an easy task. It is always useful to have some kind of protocol to guide you through the event. Put together a team or committee consisting of management and employees to regularly brainstorm worst-case scenarios. Plan disaster drills to keep employees informed and trained in crisis management skills. Designate a spokesperson responsible for talking to the media, should they become involved. Keep a notebook with instructions on how to proceed in emergency situations.

> If a foodborne illness is suspected, the first action is to remove all suspected product from sale.

Bioterrorism and Food Protection

On September 11, 2001, terrorism in the United States became a reality. As a result, the federal government and state agencies have developed plans to react to potential efforts of sabotage. Reacting in a rational manner is extremely important. There are several actions that can be used to manage these types of problems. These include the following:

The use of photo identification name tags helps to identify employees.

◆ Monitor employees and allow only approved employees into production areas. The use of photo identification name tags helps to identify employees.

◆ Protect food preparation areas from anyone who is not assigned to that area.

◆ Review visitor policies and change rules to keep areas secure. Use sign-in and sign-out logs.

◆ Prohibit personal items like lunch containers, cases, purses, and other such items from processing areas. Provide storage for employee property in a separate room or locker.

◆ Report any unusual or suspicious activity to your supervisor or manager.

◆ If a suspicious problem is identified, call the Federal Bureau of Investigation (FBI) and FDA Office of Crime Investigation. These agencies are best equipped to handle emergency situations and will advise you of the steps to follow.

Provide storage areas for employee property.

◆ Designate a spokesperson to deal with media or other inquiries. This method assures information is managed carefully and mixed messages are avoided.

Product tampering is also a food safety concern, and many food establishments have developed policies and procedures to deal with such problems. Managers should teach employees to be alert and to report any unusual activity immediately. It is always better to be overly cautious than to ignore a potential incident.

Challenges of Food Safety Management

There are a tremendous number of challenges related to managing food safety in food establishments. The foodservice and retail food industries have a high employee turnover rate and there are language and cultural barriers that make food safety education difficult. As you develop a food safety management program for your establishment, use the following guidelines:

◆ **Focus on risk:** Inspection procedures and audits for food safety management programs should focus directly on controlling risk factors that affect food safety.

◆ **Make it manageable:** The food management program needs to be user friendly. It should be easy to implement and manage. Food safety management programs need to be flexible and related to specific operational procedures. One size does not fit all.

◆ **Strengthen communication:** The foodservice and retail food industries and regulatory professionals share a common goal to protect public health by only allowing the production and sale of safe food. With the same goal, the food industry and regulatory professionals should partner and work together.

Education and training need to be encouraged for members of the food industry and for regulators. Education and training are the most important keys to effective food safety management and success of the HACCP program. Integrate your food safety management system into each food employee's duties, performance plans, and goals.

The content of the training program should provide employees with an overview of the HACCP system and how it works to ensure food safety. The primary goal of your HACCP training program is to provide employees the skills they will need when performing specific tasks (monitoring and recording) which are required by the HACCP plan. Motivate employees by stressing the importance of their roles and their responsibility to the success of the HACCP program.

Effective training and supervision will help you achieve the benefits of an HACCP- based operation. This will save you money and, more importantly, enhance the safety of the products you are serving to your clients.

What Do You Think?

Health inspector claims that state fair is not measuring up

The most popular dish at the summer state fair is a barbeque boneless pork chop sandwich. It has been a local favorite for years and it is not uncommon for lines to exceed 100 people waiting for the summer treat. But, this year the sandwich was not so popular for some. Over 700 people that attended the fair last August were diagnosed with *Salmonella* foodborne infections. Most of them had enjoyed the pork chop sandwich.

As the health inspectors studied the problem, they found that temperature-measuring devices were not used frequently to verify either cooking temperatures or hot-holding temperatures. No temperature records were kept. When the health inspector asked for the bi-metallic thermometer to be calibrated, the temperature in an ice water bath measured over 55°F (13°C), rather than the anticipated 32°F (0°C).

Summary

Back to the Story . . . When you create an effective food safety system, you can avoid situations like the one described in the case at the beginning of this chapter. A HACCP recipe and flow chart could have been implemented to alert food employees of potential hazards and critical control points to prevent, minimize, or eliminate them.

Pork products have frequently been involved as the source of foodborne salmonellosis. Proper temperatures should have been identified for the pork sandwiches to control *Salmonella* for both cooking and hot-holding. Temperatures should have been checked periodically to assure that cooking and hot-holding were being performed correctly. Temperature-measuring devices were also not functioning properly. They should be calibrated and then sanitized prior to use. Records were also not kept, which made the foodborne illness investigation more difficult. Monitoring and record keeping could have also been an important step in prevention. This is a good example of where the food safety system failed. An effective monitoring and record-keeping practice was not in place, equipment was not working properly, and the staff was not properly trained. This is an equation for disaster.

The Hazard Analysis Critical Control Point (HACCP) system is a prevention-based safety program that identifies and monitors the hazards associated with food production. The system, when properly applied, can be used to effectively control any area or point in the food flow that could cause a hazardous situation.

Food employees and managers share the responsibility for maintaining an effective record-keeping system. Food employees should measure and record all appropriate monitoring data in HACCP records, and managers should review these records to make certain that monitoring is being done properly. Well-organized monitoring records provide evidence that food safety assurance is being accomplished according to HACCP principles.

The HACCP system is the most effective method created to date to ensure the safety of food-processing and preparation operations. Implementation of a properly designed HACCP program will protect public health well beyond anything that could be accomplished using old style methods. Traditional inspections emphasized facility, equipment, design, and compliance with basic sanitation principles. HACCP focuses on the actual safety of the product.

Food establishments should also have good policies and practices in place to respond to crisis management situations that could have an impact on food safety. Be sure that a plan is in place and employees are well trained to respond appropriately.

Quiz 8 (Multiple Choice)

Please choose the BEST answer to the questions.

1. A hazard, as used in connection with a HACCP system, is:

    a. Any biological, chemical, or physical property that can cause an unacceptable risk.

    b. Any single step at which contamination could occur.

    c. An estimate of the likely occurrence of a hazard.

    d. A point at which loss of control may result in an unacceptable health risk.

2. A risk, as used in connection with a HACCP, system is:

    a. Failure to meet a required critical limit for a critical control point.

    b. An estimate of the likely occurrence of a hazard.

    c. Greatest for non-PHF (TCS).

    d. Lowest when complex recipes are required to produce a food item.

3. Which of the following statements about HACCP programs is **false**?

    a. The HACCP system attempts to anticipate problems before they happen and establish
    procedures to reduce the risk of foodborne illness.

    b. The HACCP system targets the production of PHF (TCS) for the entire flow of food.

    c. Records generated by the HACCP system can be used to aid in foodborne disease investigations.

    d. A HACCP system should only be implemented by public health officials who have been certified by the FDA to conduct such programs.

4. Which of the following statements is **false**?

    a. A critical limit is the level that must be met to ensure each critical control point eliminates a microbiological, chemical, or physical hazard.

    b. There must be at least two critical control points in the flow of food in order for a HACCP system to be implemented.

    c. Many steps in food production are considered control points, but only a few qualify as critical control points.

    d. A critical control point is a point, step, or procedure in food preparation where controls can be applied and a food safety hazard can be prevented, eliminated, or reduced to acceptable levels.

5. The ultimate success of a HACCP program depends on:

    a. Eliminating all PHF (TCS) from your operation.

    b. Providing proper training and equipment for employees who are implementing the HACCP system.

    c. Having HACCP flow charts developed for every single food sold by the establishment.

    d. Food establishment managers having sole authority for implementing the HACCP system.

6. Which of the following is an example of a critical control point?

    a. Receiving poultry and eggs only from approved sources.

    b. Rotisserie chicken is heated in the oven until the thickest part of the product reaches 165°F (74°C) for at least 15 seconds.

    c. The identified spokesperson of the retail food establishment that speaks to the media.

    d. The cutting board is washed and sanitized between chopping carrots and celery for the garden salad.

7. Which of the following programs would be considered a prerequisite program for the HACCP system?

    a. Good hygiene practices

    b. Effective cleaning and sanitizing programs

    c. Equipment maintenance programs.

    d. All of the above.

8. For which of the following products would it **not** be necessary to develop a HACCP flow chart?

    a. Chicken salad.

    b. Tuna salad.

    c. Seafood salad.

    d. Citrus fruit salad.

9. If a case of foodborne illness is suspected at your facility, you should:

    a. Isolate and label the suspected food so that it will not be sold.

    b. Keep good records of all communications with the customer(s).

    c. Not try to diagnose the problem or admit fault.

    d. All of the above.

Answers to the multiple-choice questions are provided in **Appendix A**.

References/Suggested Readings

Bryan, Frank L. 1990. *"Hazard Analysis Critical Control Point (HACCP) Systems for Retail Food and Restaurant Operations."* Journal of Food Protection 53(11): 978-983.

Bryan, F. L. 1992. *"Hazard Analysis Critical Control Point Evaluations: A Guide To Identifying Hazards and Assessing Risks Associated with Food Preparation and Storage."* World Health Organization, Geneva, Switzerland.

Bryan, F.L., C.A. Bartleson, C.O. Cook, P. Fisher, J.J. Guzewich, B.J. Humm, R.C. Swanson, and E.C.D. Todd. 1991. *Procedures to Implement the Hazard Analysis Critical Control Point System.* International Association of Milk, Food and Environmental Sanitarians, Ames, IA, 72 pp.

Code of Federal Regulations. 2001. *21 CFR 129 - Processing and Bottling of Bottled Drinking Water.* U.S. Government Printing Office, Washington, D.C.

Corlett, D.A. and Pierson, M.D. 1992. *HACCP Principles and Applications,* Van Nostrand Reinhold, New York.

Environmental Protection Agency. 2006. *Emergency Disinfection of Drinking Water.* http://www.epa.gov/safewater/faq/emerg.html.

Federal Register, Vol. 59, No. 149. U.S. Government Printing Office, 1994, Washington, D.C.

Food and Drug Administration. 2009. *2009 Food Code.* U.S. Public Health Service, Washington, D.C.

LaVella, B. and J. L. Bostic. 1994. HACCP for Food Service—Recipe Manual and Guide. LaVella Food Specialists, St. Louis, MO.

Lotkin. J. K. 1995. The HACCP Food Safety Manual. Wiley Press, New York, NY.

McSwane, D., Rue, N., Linton, R. 2005. *Essentials of Food Safety and Sanitation, 4th ed.* Prentice Hall, Upper Saddle River, NJ.

Pierson, M. D. and D. A. Corlett, Jr. 1992. *HACCP Principles and Applications.* Van Nostrand Rheinhold, New York, NY.

Pisciella, J.A. 1991. A *handbook for the practical application of the hazard analysis critical control point approach to food service establishment inspection.* Central Atlantic States Association of Food and Drug Officials, c/o William Kinder, Pennsylvania Department of Agriculture, PO Box 300, Creamery, PA 19430.

Price, R.J, P.D. Tom, and K.E. Stevenson. 1993. *Ensuring Food Safety - The HACCP Way.* University of California, Food Science & Technology Department, Davis, CA.

U.S. Food and Drug Administration. Managing Food Safety: A HACCP Principles Guide for Operators of Food Establishments at the Retail Level: Regulatory Applications in Retail Food Establishments. Center for Food Safety and Applied Nutrition, 200 C Street, SW, Washington, D.C.

Suggested Web Sites

Canadian Food Safety Sites Involving HACCP

foodnet.fit.ca/safety/safety.html

International Food Information Council

     ificinfo.health.org/

NSF International

     www.nsf.org

The Food Marketing Institute

     www.fmi.org

The National Restaurant Association

     www.nra.org

Gateway to Government Food Safety Information

     www.foodsafety.gov

U.S. Department of Agriculture (USDA)

     www.usda.gov

Centers for Disease Control and Prevention (CDC)

     www.cdc.gov

Food and Drug Administration (FDA)

     www.fda.gov

USDA Food Safety and Inspection Service (FSIS)

     www.fsis.usda.gov

USDA/FDA Food and Nutrition Information Center

     www.nal.usda.gov/fnic

Partnerships for Food Safety Education

     www.fightbac.org

Food and Agriculture Organization

     www.fao.org

Food Safety Regulations

Key Terms

- Adulteration
- Centers for Disease Control and Prevention (CDC)
- Conference for Food Protection (CFP)
- Consumer advisory
- Environmental Protection Agency (EPA)
- Food and Drug Administration (FDA)
- Generally Recognized As Safe (GRAS) substances
- Grading
- Inspection for wholesomeness
- State, local, and tribal agencies
- U.S. Department of Agriculture (USDA)

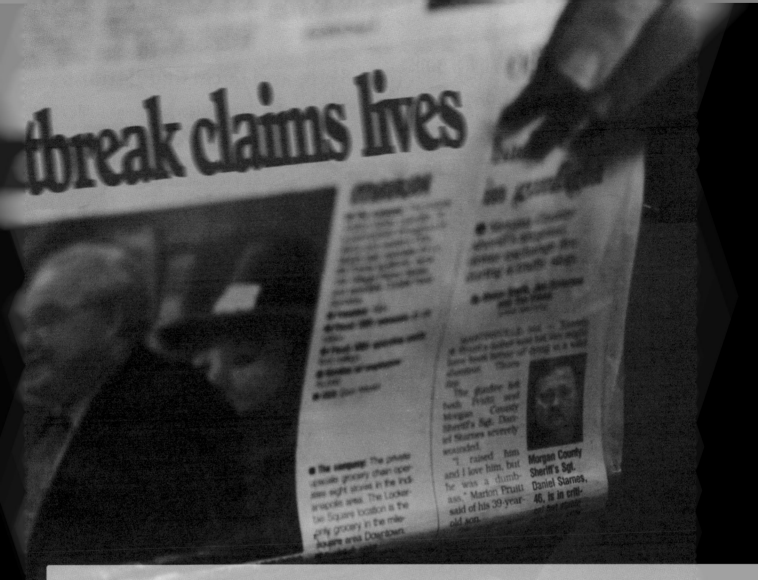

Food safety is regulated by federal, state, local, and tribal agencies.
The federal government has set up agencies to regulate the production, manufacturing, sale, and transportation of food products. It also sets standards for food grading, inspecting quality and food labeling. State, local and tribal government create laws for food safety and enforce them through permits and food inspections. This chapter focuses on the different roles of government in the food safety process.

Learning Points:

◆ Recognize the role of federal, state, local, and tribal jurisdictions in regulating and monitoring food safety for food establishments.

◆ Identify the areas of food establishments that are included in inspections and audits.

◆ Use references for federal, state, and local authorities.

◆ Identify organizations important to the food industry related to food safety.

State and Local Regulations

State, local and tribal agencies responsible for enforcing food and safety requirements may be under the department of environmental health, public health, or agriculture. Jurisdiction and laws vary from state to state. Contact your local authorities to find out what agencies are responsible for enforcing food safety regulations in your area.

Working with state and local agencies:

◆ Get a copy of the current food safety code from your local health department.

◆ Your state and local codes contain the food safety standards and regulations that apply to your type of operation.

◆ Familiarize yourself with the key food safety provisions of the code. The *Food Code* is a set of recommendations of safe food-handling practices, but the *Food Code* is not law. You will need to follow the regulations specific to the state/local food code in your area.

◆ Most local food safety agencies require a permit to operate and enforce regulations by performing routine inspections.

Permit to Operate

It is unlawful to operate a food establishment without a valid permit issued by the regulatory authority in your area. When a permit is obtained, post it in a prominent location. Permits are generally not transferable from one person to another, from one food establishment to another, or from one food operation to another. If you add a partner or incorporate, these actions are considered ownership changes and may require you to apply for a new permit to operate.

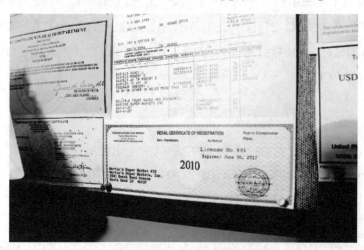

Post your permit to operate in a prominent place in your establishment.

Routine Inspections

Food establishments are routinely inspected by regulatory agencies to assure they are complying with the food safety rules and regulations of the jurisdiction.

Always cooperate with inspecting personnel. Greet the inspector when he or she arrives and ask to see their official identification.

Food managers should periodically conduct self-inspections of their establishment to assure proper sanitation and food safety. During a self-inspection, the manager should look for signs of contamination, improper food-handling practices, and other conditions that may put food safety in jeopardy.

How Often Will You Be Inspected?

Many regulatory agencies use a risk-based approach for setting inspection frequencies. Under this system, the risk a food establishment poses is calculated using such factors as:

◆ The sanitation history of the establishment
◆ Number of home meal replacements sold
◆ The amount of food preparation and handling conducted on-site
◆ Condition of building
◆ Critical violations observed.

Phases in a Routine Inspection:

A routine inspection or audit normally consists of three phases:

1. A **pre-inspection conference** where the inspector and manager, or person in charge, review previous inspection results and discuss information relevant to the current inspection.

2. The **current inspection** is conducted.

3. A **post-inspection conference** is conducted where the results of the current inspection are reviewed and discussed with the manager or person in charge.

Most Inspections Focus On:

◆ The condition of food ingredients, foods, and supplies
◆ Personal hygiene and employee health practices
◆ Temperatures of food and food-holding and food-storage equipment
◆ Cleaning and sanitizing procedures
◆ Condition of equipment and utensils
◆ Water supply and waste disposal procedures
◆ Pest control
◆ Any other aspect of the operation that might compromise food safety.

Inspection Provisions

In the 2009 version of the *Food Code*, FDA has designated a 3-tiered set of definitions to rank risk. The three new terms are:

According to Food Code:

The criteria for inspections have changed in the 2009 version of the *Food Code*. Violations in previous versions of the *Food Code* were designated as "critical" or "non-critical" violations.

In the 2009 version of the *Food Code*, the FDA has designated a 3-tiered set of definitions to rank risk. The three new terms are:
◆ **Priority item**
◆ **Priority foundation item**
◆ **Core item**

Priority Item

A provision of the *Food Code* whose application contributes directly to the elimination, prevention, or reduction of foodborne hazards to levels that will not cause harm or injury.
◆ **Immediate action must be taken to correct the violation of these items**
◆ Priority Foundation Items include quantifiable measures to show control of hazards such as cooking, cooling, reheating, and hand washing.

Priority Foundation Item

Linked to a Priority Item and it supports, enables, or helps achieve it. Active managerial control/industry control systems are used to support the compliance of Priority Items.

◆ Conducting personnel training
◆ Monitoring and enforcing Priority Item activities
◆ Providing necessary equipment, facilities, etc., to carry out Priority Item activities
◆ Developing & carrying out HACCP plans when necessary,
◆ Maintaining documents or records, as necessary
◆ Labeling food for employees or consumers.

Core Item

A good retail practice (GRP) which is not intended to control a particular hazard but all hazards in general.

Core Items focus on:
◆ General sanitation
◆ Sanitation SOPs
◆ Equipment design
◆ Design & construction of facilities and structures
◆ General maintenance
◆ Operational controls.

Federal Agencies

The primary federal agencies that protect our food supply and their functions are provided in the following charts.

 **U.S. Food and Drug Administration (FDA)**

- Regulates the processing, manufacturing, and interstate sale of many food items, except for meat, poultry, and egg products.
- Sets standards with respect to composition, quality, labeling, and safety of foods and food additives.
- Publishes the *Food Code.*
- Protects the public's health by preventing the adulteration and misbranding of food.
- Assures food shipped in interstate commerce is safe, pure, wholesome, sanitary, and honestly packaged and labeled.
- Maintains a list of interstate certified shellfish shippers and a list of interstate milk shippers.
- The FDA Office of Criminal Investigations investigates suspected criminal violations of the Federal Food, Drug, and Cosmetic Act, the Federal Anti-tampering Act, and other related federal statutes. These investigations concentrate on significant violations of these laws, with a priority on conduct that may present a danger to the public health.

 U.S. Department of Agriculture (USDA)

- Inspects domestic and imported meats, poultry, eggs, catfish, and processed meat and poultry products. USDA maintains a list of approved facilities for meat and poultry processing.
- Conducts voluntary grading services for red meats, poultry and eggs, dairy products, and fruits and vegetables.

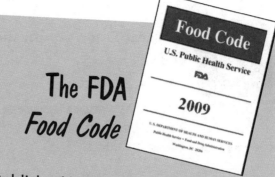 **The FDA Food Code**

- Published by the FDA, the *Food Code* is the model for food programs across the country. **It is not a law itself, but federal, state, local, and tribal agencies create laws for food safety based on the guidelines and recommendations in the *Food Code*.**
- The *Food Code* is updated every 2 years based on the recommendations made by the Conference for Food Protection (CFP).

 U.S. Department of Commerce (USDC)

- Develops grade standards for processed fishery products.

 National Marine Fisheries Service (NMFS) and NOAA Seafood Inspection Program (SIP)

- Division of the USDC.
- Provides voluntary inspection service for processed fishery products.
- Maintains a list of approved fish processors and fishery products on a permanent basis.

Participating establishments contract with the Seafood Inspection Program for sanitation inspection services and have been inspected, approved, and certified by the program as being capable of producing safe, wholesome products in accordance with specific quality regulations enacted by the Department of Commerce.

Federal Trade Commission (FTC)

◆ Involved in enforcing various laws regarding marketing practices and national advertising of foods and other products.

Occupational Safety and Health Administration (OSHA)

◆ Created in the U.S. Department of Labor to enforce the Occupational Safety and Health Act. It enforces health standards and regulations on safety, noise, and other workplace-related hazards.

Centers for Disease Control and Prevention (CDC)

◆ Responsible for protecting public health through the prevention and control of diseases
◆ Supports foodborne disease investigations and prepares annual summaries and statistics on outbreaks of diseases transmitted through food and water.

Environmental Protection Agency (EPA)

◆ Created to prevent, control, and reduce air, land, and water pollution
◆ Regulates the use of toxic substances (pesticides, sanitizers, and other chemicals), monitors compliance, and provides technical assistance to states.

Other Food Safety Related Organizations

The Conference for Food Protection (CFP)

The Conference for Food Protection (CFP) is a nonprofit organization that represents the major stakeholders in retail food safety. Its membership includes:

◆ Representatives of the food industry
◆ Government
◆ Academic community
◆ Professional organizations
◆ Consumer groups.

The CFP membership meets at least every other year to identify food safety problems, make recommendations, and implement practices to assure food safety. Several committees are active in-between meeting times. The CFP has no formal regulatory authority, yet over the years it has been able to effectively influence the content of model food laws and regulations.

Other organizations also help protect our food supply. Many of the organizations listed below are good sources for information and training materials:

◆ The American Public Health Association (APHA)
◆ The Association of Food and Drug Officials (AFDO)
◆ The Frozen Food Industry Coordinating Committee (FFICC)
◆ The Food Research Institute (FRI)

- Institute of Food Technologists (IFT)
- Grocery Manufacturers Association (GMA)
- National Conference on Interstate Milk Shipments (NCIMS)
- International Association for Food Protection (IAFP)
- The National Environmental Health Association (NEHA)
- The National Pest Management Association (NPMA)
- The National Restaurant Association (NRA)
- NSF International (NSF)
- The National Shellfish Sanitation Program (NSSP)
- Underwriters Laboratories, Inc. (UL)
- World Health Organization (WHO).

> **There are many organizations that help protect our food supply.**

Inspection for Wholesomeness and Grading of Food

Federal agencies play an important role in the inspection and grading of our food supply. **Inspection for wholesomeness** refers to an examination of domestic and imported meat, poultry, and egg products to assure they are wholesome and free from adulteration. These inspections are required by law and are conducted by veterinarians and technicians who work for the Food Safety and Inspection Service (FSIS) arm of USDA. The USDA can also be involved with microbial testing of foods.

E. coli O157:H7 has been declared an adulterant in raw ground beef, which means that no detectable amount of these bacteria is permitted in ground beef by law. A national microbial sampling plan has been implemented to determine the prevalence of *E. coli O157:H7* bacteria in ground beef. Under the sampling plan, USDA officials take samples of ground beef from federally inspected plants and food establishments and test them for *E. coli O157:H7* bacteria. If a sample is confirmed positive for *E. coli 0157:H7,* the food establishment may be asked to recall the product and communicate this information to the public.

USDA also has similar testing programs for *Salmonella* bacteria in raw and ready-to-eat foods and for *Listeria monocytogenes* in ready-to-eat meats.

Grading refers to the process of evaluating foods relative to specific, defined standards in order to assess its quality. Grading is a voluntary activity. However, most food processors participate in grading programs because their customers prefer to buy products that have met specific quality standards. More information on grading is provided in Chapter 4 of this book.

Substances used in foods for years and with no apparent ill effects are commonly identified as **Generally Recognized As Safe (GRAS)**. Examples of substances that fall into the GRAS category are spices, natural seasonings, flavoring materials, fruit and beverage acids, baking powder chemicals, and drying agents. The purpose of establishing a GRAS list is to recognize the safety of basic substances without the requirement for rigorous safety testing.

Food Labeling

Governmental agencies such as FDA and USDA have labeling requirements to assure that consumers are given complete, accurate and useful information about the food products they purchase. The USDA oversees labeling of meat,

KEY TERMS

Inspection for wholesomeness - an examination of domestic and imported meat, poultry, and egg products to assure they are wholesome and free from adulteration.

Grading - the process of evaluating foods relative to specific, defined standards in order to assess its quality.

Generally Recognized As Safe (GRAS) - GRAS stands for generally recognized as safe. These are substances added to foods that have been shown to be safe based on a long history of common usage in food.

poultry, and egg products, and FDA is responsible for assuring that other types of foods sold in the United States are properly labeled. This applies to foods produced domestically, as well as foods from foreign countries.

Label information is vital for both economic and health reasons. Truthful and complete information on the label permits consumers to make value comparisons among competing products. The Federal Food, Drug, and Cosmetic Act and the Fair Packaging and Labeling Act are the federal laws governing food products under FDA's jurisdiction. In addition, the Nutrition Labeling and Education Act requires most foods to bear nutrition labeling, and requires food labels that bear nutrient content claims and health messages to comply with specific requirements.

People look at food labels for different reasons. But whatever the reason, consumers want this information to be simple, accurate, and easy to understand.

Safe Food-Handling Label

A safe food-handling instruction label provides helpful food-handling information for the consumer on all raw or partially precooked (not ready-to-eat) meat and poultry. The label was designed to educate the consumer for storage, preparation, and handling of raw meat and poultry products in the home.

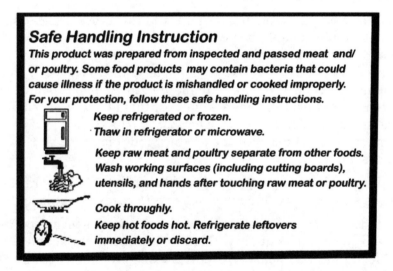

USDA safe-handling instruction label
(Source: http://www.usda.gov/agency/oc/design/art_symbols.html)

Food Allergen Labeling and Consumer Protection Act

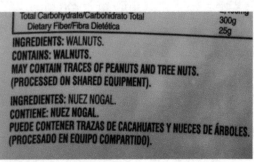

Allergen warning with a "contains" statement.

The Food Allergen Labeling and Consumer Protection Act (FALCPA) mandates labeling requirement for foods that contain one or more of the eight major food allergens—peanuts, eggs, milk, wheat, tree nuts, soybeans, fish, and crustacean shellfish—or a food ingredient that contains protein derived from one of these foods (see Chapter 2).

Allergens can be identified in either of two ways. One option is to use a "contains" statement, which means that the word *contains* is followed by a list of all the major food allergens contained in the product. The second option is "parenthetical" listing, which means that within the list of ingredients, the common or usual name of the food allergen is included in parentheses following the ingredient name.

The person in charge (PIC) of a food establishment must be able to:

◆ Demonstrate knowledge about the eight Major Food Allergens and the symptoms they cause.

◆ Identify the allergens and their related symptoms when asked by an inspector.

◆ Know what actions to take if an allergic reaction occurs in his or her store.

◆ Assure employees are properly trained in food safety, including food allergy awareness.

Product Dating Labels

A product dating label is a date stamped on a food package to help food establishments that sell these types of products (i.e., supermarkets) know how long to display the product for sale. It can also help customers understand the time limit to purchase or use the product at its best quality. Date labels are on almost all food products. Date labels are usually found on perishable PHF (TCS) such as meat, poultry, eggs, and dairy items.

There are many types of dating labels that can be used. These labels were introduced briefly in Chapter 8 of this book.

◆ A "sell-by" date label tells the food establishment how long to display the product for sale. Food products should be removed from the shelves immediately after the sell-by date, and customers should buy the product before the date expires.

◆ A "best if used by" (or before) date is used to indicate the time period recommended for the best flavor or quality. A "use-by" date is the last date recommended for the use of the product while at peak quality. The manufacturer of the product usually determines the date by using quality control and shelf-life tests. Federal laws regulate date marking of infant formula products.

Sell-by date

Date Marking for Ready-to-Eat, PHF (TCS)

Product date marking is used by food establishments to indicate the date or day by which refrigerated, ready-to-eat, PHF(TCS) must be consumed, sold, or discarded. There are date marking provisions that are included in the *Food Code* for the following situations:

1. Food establishments that prepare and hold refrigerated, ready-to-eat, PHF(TCS) for more than 24 hours.

   a. Food establishment personnel must clearly mark these products to indicate the date or day by which the food must be consumed on the premises, sold, or discarded using the temperature and time combination of 41°F (5°C) or less for a maximum of seven days. The day of preparation will be counted as Day 1.

   b. This date marking requirement does **not** apply to reduced oxygen packaged (ROP) foods prepared and held in food establishments. However, special handling and storage are required for ROP foods as described in Chapter 4 of this book.

2. Food establishments that serve and sell refrigerated, ready-to-eat PHF (TCS) prepared and packaged by an inspected food processing plant.

Date marking indicates the date by which the food must be used.

a. If the food is to be held for more than 24 hours, the product must be clearly marked at the time the original container is first opened in a food establishment to indicate the date or day by which the food shall be consumed on the premises, sold, or discarded, based on the temperature and time combinations noted previously. The day the original container is first opened in the food establishment shall be counted as Day 1, and the day or date marked by the food establishment may not exceed a manufacturer's use-by date if the manufacturer determined the use-by date based on food safety.

b. A refrigerated, ready-to-eat, PHF (TCS) ingredient or a portion of a refrigerated, ready-to-eat, PHF (TCS) that is subsequently combined with additional ingredients or portions of food shall retain the date marking of the earliest prepared or first prepared ingredient.

The requirements described previously do **not** apply to:

1. Individual meal portions served or repackaged for sale from a bulk container upon a consumer's request.

2. Certain types of foods that pose very low or low risk of contamination and growth of *Listeria monocytogenes*. A listing of these exempted foods is provided at the end of this section.

Two examples of date marking systems that meet the criteria prescribed in the *Food Code* include:

1. Marking the date or day the food is prepared or the original container is first opened in a food establishment, with a procedure to discard the food on or before the last date or day by which the food must be consumed on the premises, sold, or discarded.

2. Using calendar dates, days of the week, color-coded marks, or other effective marking methods provided that the marking system is disclosed to the regulatory authority upon request. Based on the results of numerous *Listeria monocytogenes* health risk assessments and recommendations from the 2004 CFP meeting, the FDA has reevaluated its date marking rules in order to place the greatest emphasis on very high- and high-risk foods, while exempting foods that pose very low- or low-risk of contamination and growth of *Listeria monocytogenes*.

Based on this evaluation, the FDA has determined that date marking provisions do not apply to the following foods prepared and packaged by a food processing plant inspected by a regulatory authority:

Date marking labels

(Courtesy of Ecolab Food Safety Specialties)

1. Deli salads, such as ham salad, seafood salad, chicken salad, egg salad, pasta salad, potato salad, and macaroni salad, manufactured in accordance with *21 CFR 110—Current Good Manufacturing Practice in Manufacturing.*

2. Hard cheeses containing not more than 39% moisture as defined in *21 CFR 133—Cheeses and Related Cheese Products,* such as cheddar and parmesan.

3. Semi-soft cheeses containing more than 39% moisture, but not more than 50% moisture, as defined in *21 CFR 133—Cheeses and Related Cheese Products,* such as blue, gorgonzola, and Monterey jack.

4. Cultured dairy products as defined in *21 CFR 131—Milk and Cream,* such as yogurt, sour cream, and buttermilk.

5. Preserved fish products, such as pickled herring and dried or salted cod, and other acidified fish products defined in *21 CFR 114—Acidified Foods.*

6. Shelf stable, dry fermented sausages, such as pepperoni and Genoa salami that are not labeled "keep refrigerated" as specified in *9 CFR 317—Labeling, Marking Devices, and Containers,* and that retain the original casing on the product.

7. Shelf stable salt-cured products such as prosciutto and Parma (ham) that are not labeled "keep refrigerated" as specified in *9 CFR 317—Labeling, Marking Devices, and Containers.*

Foods that do not have to be date marked.

Consumer Advisory

Food establishments that sell or serve raw or undercooked animal foods or ingredients for human consumption must inform their customers about the increased risk associated with eating these kinds of foods. The disclosure is used to make customers aware of the dangers of raw or undercooked foods and of items that either contain or may contain raw or undercooked ingredients. The reminder is a notice about the relationship between thorough cooking and food safety.

Disclosure is satisfied when items are described such as oysters on the half-shell (raw oysters), and raw-egg Caesar salad; or items are asterisked to a footnote that states the items:

◆ Are served raw or undercooked, or;

◆ Contain (or may contain) raw or undercooked ingredients.

The reminder is satisfied when the items requiring disclosure are asterisked to a footnote that states information about:

◆ The safety of these items, written information is available upon request

◆ Eating raw or undercooked meats, poultry, seafood, shellfish, or eggs, which may increase your risk of foodborne illness

A food establishment can satisfy the consumer advisory requirements by providing both a disclosure and a reminder.

:oup   Chicken breast 5.50   carrots, celery 8

Chips, guacam

Asian chicken salad with Thai chili lime vinaigrette
grilled chicken, cucumbers, carrots, scallions and peanuts 7.75

"Consuming raw or under-cooked meats, poultry, seafood, shellfish or eggs may increase your risk of foodborne illness."

◆ Increased risk of foodborne illness when the above products are eaten if you have certain medical conditions.

Disclosure of raw or undercooked animal-derived foods or ingredients and reminders about the risk of consuming such foods should appear at the point where the customer selects the food. Both the disclosure and the reminder need to accompany the information from which the customer makes a selection. That information could appear in many forms such as display signage, package labels, and consumer brochures.

Consumer advisories may be tailored to be product specific if a food establishment either has a limited menu or offers only certain animal-derived foods in a raw or undercooked, ready-to-eat form. For example, a raw bar serving molluscan shellfish on the half-shell, but no other raw or undercooked animal food, could elect to confine its consumer advisory to shellfish.

Information about Food Safety Guidelines and Regulations

Government organizations work very hard to promote safe food and to communicate to food establishment employees and consumers alike the importance of safe food practices. There are many consumer initiatives and media campaigns designed to educate people to the issues surrounding food safety.

The Fight BAC!™ Campaign

The Fight Bac!™ Campaign is an educational program developed by industry, government, and consumer groups to teach consumers about safe food handling. The focal point of the campaign is BAC (short for bacteria), a character used to make people aware of the invisible germs that cause foodborne illness. Even though people cannot see, smell, or taste BAC, it and millions more germs like it are in and on food and food-contact surfaces. The Fight Bac!™ Campaign uses media and community outreach programs to teach consumers how to keep food safe from harmful bacteria through 4 simple steps:

1) wash hands and surfaces often,

2) prevent cross contamination,

3) cook foods to proper temperature, and

4) refrigerate foods quickly.

The partners in the Fight Bac!™ Campaign use newspaper articles, public service announcements, printed brochures, and school curricula to teach consumers how to lower their risk of foodborne illness.

Be Food Safe

In 2007, the Partnership for Food Safety Education launched the Be Food Safe program, aimed at reminding consumers about the important safe food-handling practices.

The four easy lessons of Clean, Separate, Cook, and Chill can help consumers prevent harmful bacteria from making people sick.

This information is especially important for the caregivers of people most vulnerable to foodborne illnesses, including children, seniors, and people with compromised immune systems (http://www.befoodsafe.org/).

Additional Information Web sites

With the new electronic age that we are all experiencing, there are many good websites that provide information on food safety and food safety regulations. Some of the more commonly used sites that are helpful for individuals that work in food establishments are listed below:

◆ FDA *Food Code* (http://www.fda.gov/Food/FoodSafety/RetailFoodProtection/FoodCode/FoodCode2009/default.htm)

Provides a copy of the 2009 *Food Code*.

◆ Gateway to Federal Food Safety Information (http://www.foodsafety.gov/)

Designed to provide information on meat, poultry and fish, dairy and egg products, fruits and vegetables, foodborne illness and emerging issues in food safety.

◆ FDA Product Specific Information (http://www.fda.gov/Food/FoodSafety/Product-SpecificInformation/default.htm)

An FDA site that provides regulatory information about acidified and low-acid canned foods, bottled water and carbonated soft drinks, cheese safety, egg safety, fruits, vegetables and juices, infant formula, medical foods, milk safety, seafood.

◆ USDA-FSIS Regulations (http://www.fsis.usda.gov/Regulations_&_Policies/index.asp)

Provides information about regulations for meat, poultry, and egg products.

◆ Retail Food Safety Consortium (www.retailfoodsafety.org)

A new web site that provides regulatory information about state and local jurisdictions throughout the United States. It is also a good resource for additional training materials.

◆ Conference for Food Protection (www.foodprotect.org)

Provides invaluable information about emerging issues related to retail food safety as they impact the development and change in the *Food Code*.

Food safety regulations

A large multistate recall has just been initiated for leafy greens (lettuce, spinach, and salad mixes) because of the possible presence of the *E. coli* O157:H7 bacteria. The recall covers products from three different brands (Sally's Salads, Murphy's Fresh, and Seasons) with the following codes: AMP6134, AMP6135, AMP6136, AMP6137, and AMP6138.

Restaurant owners, supermarkets, and other retailers are asked to remove all suspected products from storage and display areas. Retail food stores are also encouraged to set up a procedure for communicating with customers that may have purchased or consumed affected products. Consumers who have purchased the leafy greens, but have not consumed the product, are urged to return the product to the place of purchase or call the company's toll-free number for a full refund.

Summary

Back to the Story...

The case presented at the beginning of the chapter provides an illustration of a food recall that was established by 3 different food processors that produce fresh leafy greens. The recall is a result of confirmed illnesses and product testing which revealed that some of the finished product contained the bacteria *Escherichia coli* O157:H7. Whenever a company suspects or confirms that something may be wrong with one of its products, it will issue a recall to assure the product is removed from store shelves and that customers who may have already purchased or consumed the products will have awareness of the recalled product. Regulatory agencies will frequently participate in recalls to assure suspect food products are removed from sale as quickly as possible. It is important that you have a process in place and a practiced plan to deal with food recalls. You will need to understand how to identify product codes on food ingredients that you use, develop a system to remove these products from your operation, and, effectively communicate recall related information to your customers.

Federal, state, and local food safety regulations are important for maintaining a safe food supply. State, local, and tribal agencies are the most critical point of contact when it comes to monitoring and enforcing each state or local food code. They evaluate food establishments to help assure delivery of safe food to customers.

It is essential that members of the food industry and local/state/federal/tribal regulatory agencies work together in an effort to protect the health of customers and meet their expectations for both food safety and quality.

Quiz 9 (Multiple Choice)

Please choose the BEST answer to the questions.

1. Which federal agency is responsible for the updates of the *Food Code*?
   a. Food and Drug Administration.
   b. Centers for Disease Control and Prevention.
   c. U.S. Department of Agriculture.
   d. Environmental Protection Agency.

2. Which federal agency is responsible for ensuring the safety of domestic meat, poultry, and egg products?

   a. Centers for Disease Control and Prevention.
   b. Environmental Protection Agency.
   c. Food and Drug Administration.
   d. U.S. Department of Agriculture.

3. Which federal agency is responsible for ensuring the safety of all foods except domestic meat, poultry, and egg products?

   a. Centers for Disease Control Prevention.
   b. Environmental Protection Agency.
   c. Food and Drug Administration.
   d. U.S. Department of Agriculture.

4. The organization, that meets every other year, to discuss and provide recommendations for development of the *Food Code* is called:

   a. The Center for Disease Control and Prevention.
   b. The Conference for Food Protection.
   c. The Food Safety Consortium.
   d. The Fight Bac™ Campaign.

5. Which of the following statements about food safety inspections is **false**?

   a. A routine inspection normally consists of pre-inspection, inspection, and post-inspection phases.
   b. Food managers should conduct periodic self-inspections to assure proper food safety practices are followed.
   c. Failure to correct violations of priority items can result in fines and possible closure of the food establishment.
   d. Food establishments that have a HACCP system in place are not routinely inspected by regulatory officials in their jurisdiction.

6. Information used to inform customers about the increased risk associated with eating raw or undercooked animal foods is called a:

   a. Public disclosure.
   b. Product date label.
   c. Consumer advisory.
   d. Recall notice.

7. Information used to help food establishment employees know how long to display a product for sale is called a:

   a. Public disclosure.
   b. Product date label.
   c. Consumer advisory.
   d. Recall notice.

8. During an inspection of a food establishment, the manager should:

   a. Ask to see credentials.
   b. Assist the inspector whenever possible.
   c. Discuss any violations.
   d. All of the above.

9. The term GRAS stands for:

   a. Good Retail Advisory Services.

   b. Ground Rules About Sanitation.

   c. Generally Recognized as Safe.

   d. None of the above.

10. Which of the following new inspection terms is used to describe practices that have the highest risk of causing foodborne illness or injury?

    a. Critical item

    b. Noncritical item.

    c. Core item.

    d. Priority item

Answers to the multiple-choice questions are provided in **Appendix A.**

References/Suggested Reading

Food Safety and Inspection Service, U.S. Department of Agriculture, Washington, D.C. 20250-3700, August 2000. http://www.fsis.gov/oa/pubs/dating.htm

IFT Publication. (1992. *Government Regulation of Food Safety: Interaction of Scientific and Societal Forces.* Food Technology. 46(1).

Labuza, T. P., and Baisier. 1992. *The Role of the Federal Government in Food Safety.* Critical Reviews in Science and Nutrition. 31(3). 167-176.

National Advisory Committee on Microbiological Criteria for Foods. (ugust 2004. "Considerations for Establishing Safety-Based Consume-By Date Labels for Refrigerated Ready-to-Eat Foods." www.fsis.usda.gov/ophs/nacmcf/2004/NACMCF_Safetybased-Date_Labels_ 082704.pdf.

Potter, N.N., and J. H. Hotchkiss. 1995. *Food Science.* Chapman and Hall, New York, NY.

Web Sites

Conference for Food Protection

  www.foodprotect.org

2009 FDA *Food Code*

  http://www.fda.gov/Food/FoodSafety/RetailFoodProtection/FoodCode/FoodCode2009/default.htm

FDA Product Specific Information

  http://www.fda.gov/Food/FoodSafety/Product-SpecificInformation/default.htm

Food and Agriculture Organization

  www.fao.org

Food Marketing Institute

  www.fmi.org

Gateway to Government Food Safety Information

www.foodsafety.gov

U.S. Department of Agriculture (USDA)

www.usda.gov

USDA-FSIS Regulations

http://www.fsis.usda.gov/Regulations_&_Policies/index.asp

Centers for Disease Control and Prevention (CDC)

www.cdc.gov

Environmental Protection Agency (EPA)

www.epa.gov

Food and Drug Administration (FDA)

www.fda.gov

National Marine Fisheries Services

www.seafood.nmfs.noaa.gov

Occupational Safety and Health Administration (OSHA)

www.osha.gov

Partnerships for Food Safety Education

www.fightbac.org

Retail Food Safety Consortium

www.retailfoodsafety.org

USDA Food Safety and Inspection Service (FSIS)

www.fsis.usda.gov

USDA/FDA Food and Nutrition Information Center

www.nal.uda.gov/fnic

World Health Organization

www.who.int

Notes

Glossary

Abrasive cleaners - cleaning compounds containing finely ground minerals used to scour articles; can scratch surfaces.

Acceptable level - within an established range of safety.

Accessible - easy to reach or enter.

Acid - a substance with a pH of less than 7.0.

Additives - natural and man-made substances added to a food for an intended purpose (such as preservatives and colors) or added unintentionally (such as pesticides and lubricants).

Adulterated - a food shall be deemed to be adulterated if it meets one of the conditions described in Section 402 of the Federal Food, Drug, and Cosmetic Act. A detailed list of conditions and exclusions associated with adulterated foods can be found at http://www.fda.gov/opacom/laws/fdcact/fdcact4.htm

Aerobe - an organism, especially a bacterium, that requires oxygen to live.

AIDS- Acquired Immune Deficiency

Air curtain - a window or doorway equipped with jets that force air out and across the opening to keep flying insects from entering.

Air gap - an unobstructed, open vertical distance through air that separates an outlet of the potable water supply from a potentially contaminated source like a drain.

Air-dry - to dry in room air after cleaning, washing, or sanitizing.

Alkaline - a substance that has a pH of more than 7.0.

Allergen - see food allergen and major food allergen.

Ambient - a recommended air temperature used to transport or to store perishable foods.

Anaerobe - an organism, especially a bacterium, that does not require oxygen or free oxygen to live.

Anisakiasis - disease caused by the *anisakis* parasite.

***Anisakis* spp.** - roundworm parasite found in fish.

Aseptic processing and packaging - a method in which food is sterilized or commercially sterilized outside the can and then aseptically placed in previously sterilized containers, which are then sealed in an aseptic environment. This method may be used for liquid foods, such as concentrated milk and soups.

Asymptomatic - without obvious symptoms or indications of a disease or other medical condition, not showing symptoms because symptoms have occurred but stopped and because symptoms never became noticeable.

Bacillus cereus - sporeforming bacterium that can live with or without oxygen and causes a foodborne illness.

Backflow - flow of contaminated water into the potable supply; caused by backpressure.

Back room - a storage area for excess products kept on hand to restock the sales floor as needed.

Backsiphonage - a form of backflow that can occur when pressure in the potable water supply drops below pressure in the flow of contaminated water.

Bacteria - single-celled microscopic organisms.

Bactericide - substance that kills bacteria.

Bacterium - one microorganism.

Bagger - a retail clerk or associate who bags customers' purchases in plastic or paper bags to suit customers.

Bagging - a process of properly, carefully packing customer's purchases in plastic or paper bags to suit customers.

Bake-off - an in-store baking process using frozen doughs and products to prepare fresh products (i.e., fresh rolls, bread, doughnuts, or other pastries).

Bale - a large bundle of cardboard that is recycled.

Baler - a device used to compact and bind corrugated cardboard into bales for recycling.

Bar code - a unique identification code on products, pallets, and coupons. The code is read by an electronic scanner for receiving, ordering, and inventory control purposes.

Bi-metallic stem thermometer - food thermometer used to measure product temperatures.

Binary fission - the process by which bacteria grow. One cell divides to form two new cells.

Biofilm - a layer of stubborn soil that develops on surfaces that are improperly cleaned or sanitized; bacteria can accumulate and grow on this surface.

Blast chiller - special refrigerated unit that quickly freezes food items.

Bloating - a damaged, swollen processed food can or glass container, which may indicate contamination, a safety hazard.

Bloom - an indication of freshness and quality as beef turns bright red when exposed to oxygen.

Bodily fluids - human secretions such as sweat, mucous, saliva, feces, and skin oil.

Bottle returns - beverage bottles returned to a retailer for recycling.

Botulism - type of food intoxication caused by *C. botulinum*.

Box cutter - a knife-like device with a razor blade used to open boxes.

Braise - to cook meat by browning it in fat, then simmering it in a covered pan with a little liquid.

Breaking down - removing products from a case in order to clean and sanitize it; also, removing component parts of a piece of equipment, such as a slicer or a grinder, to clean and sanitize it.

Bulk produce - loose, unpackaged, fresh produce that customers select themselves.

Bulk product - unpackaged, fresh products displayed in bins in large quantities and sold by the piece or the pound, such as grains, candy, and snacks.

Calibrate - to determine and verify the scale of a measuring instrument with a standard. Thermometers used in food establishments are commonly calibrated using an ice slush method (32°F or 0°C) or a boiling point method (212°F or 100°C).

Campylobacter jejuni - a microaerophilic non-sporeforming bacterium that causes a foodborne illness.

CAP - controlled atmosphere packaging.

Carrier - human, animal or insect that harbors a disease-causing microbe without having the illness and cannot work in foodservice

Case, refrigerated - a refrigerated display unit for perishable products, such as dairy products or ice cream.

Centralized prepackaging - a centralized facility that processes prepackaged perishables and ships the goods to stores.

Channel of distribution - the producers and distributors of products from a farm to the table, a path that includes a grower, producer, manufacturer, broker, wholesaler, store, and consumer.

Checker - a front-end employee who rings up, totals, and collects for a customer's order. Also known as a cashier.

Checking - the process of recording customer purchases, taking payment, making change, processing coupons, bagging, and all other functions inherent to the front-end operation.

Chemical sanitizers - products used on equipment and utensils after washing and rinsing to reduce the number of disease-causing microbes to safe levels.

Chlorine - chemical sanitizer.

Ciguatoxin - a toxin from reef-feeding fish that causes foodborne illness.

Clean - free of visible soil but not necessarily sanitized; surface must be cleaned before it can be sanitized.

Cleaning - removal of visible soil but not necessarily sanitized. A surface must be cleaned before it can be sanitized.

Cleaning agent - a chemical compound formulated to remove soil and dirt.

Clean-in-place (CIP) - equipment designed to be cleaned without moving, usually large or very heavy.

Clostridium botulinum - an anaerobic sporeforming bacterium that causes foodborne illness.

Clostridium perfringens - an anaerobic sporeforming bacterium that causes foodborne illness.

Code - a systematic collection of regulations, or statutes, and procedures designed to protect the public.

Coffin case or coffin freezer - a waist-high fixture used to display frozen food, with a transparent door or no door for easy access. See upright freezer.

Cold-holding - refers to the safe temperature range (less than 41°F or 5°C) for maintaining foods cold prior to service for consumption.

Cold storage - a facility that stores frozen foods and perishable items that need refrigeration or special handling.

Commingle - to combine shell stock harvested on different days or from different growing areas as identified on the tag or label, or to combine shucked shellfish from containers with different container codes or different shucking dates.

Comminuted - the resulting size of a food mass achieved by methods including chopping, flaking, grinding, or mincing.

Commissary - A centralized food preparation facility that distributes prepared product to stores. Also a U.S. government–subsidized, nonprofit food store, operating on a military base. Operates on a nonprofit basis and sells products at cost plus a small markup.

Compactor, trash - a device used to crush dry or wet garbage. Often found in many stores in two separate units—one compactor for paper and cardboard and one for all other materials.

Conditional employee - a potential food employee to whom a job offer is made, conditional on responses to subsequent medical questions or examinations designed to identify potential food employees who may be suffering from a disease that can be transmitted through food.

Consumer - an end user of any product or service. A shopper, patron, or customer. The final link in the chain of product distribution: manufacturing, selling, wholesaling, retailing, and consuming.

Contamination - the unintended presence of harmful substances or conditions in food that can cause illness or injury to people who eat the infected food.

Controlled atmosphere packaging (CAP) - a product in a low oxygen, nitrogen rich wrap that preserves freshness.

Convenience store - a small, easily accessed food store with a limited assortment. Many convenience stores also sell fast-food and gasoline.

Conventional supermarket - a large, self-service, retail food store with moderate pricing and selection, and annual sales in the $2- to- $8million range. Usually includes a meat, produce, dairy, and grocery department.

Cook-chill - a cooking and food preservation process for foodservice products. Food is cooked, packed, sealed, and quick chilled in a plastic pouch and stored at temperatures below 41°F (5°C).

Cooking - the act of providing sufficient heat and time to a given food to affect a change in food texture, aroma, and appearance. More importantly, cooking assures the destruction of foodborne pathogens inherent to that food.

Cooler - a refrigerated holding unit in a warehouse or store for perishables.

Cooling - the act of reducing the temperature of properly cooked food to 41°F (5°C) or below.

Cooling methods - various techniques used to promote the rapid cooling of properly cooked food to 41°F (5°C) or below. Some commonly used cooling methods include 1) placing food in shallow pans, 2) dividing large masses of food into smaller portions, 3) placing hot food in an ice water bath, 4) using cooling paddles to stir food, 5) blast chillers, 6) walk-in coolers, and, 7) adding ice as an ingredient to a condensed food.

Corrosion resistant - materials that maintain their original surface characteristics under continuous use in foodservice with normal use of cleaning compounds and sanitizing solutions.

Coving - a curved sealed edge between the floor and wall that makes cleaning easier and inhibits insect harborage.

Core item - a provision in the FDA Food Code that is not designated as a priority item or a priority foundation item. Core item includes an item that usually relates to general sanitation, operational controls, sanitation standard operating procedures (SSOPs), facilities or structures, equipment design, or general maintenance.

Critical control point (CCP) - means a point or procedure in a specific food system where loss of control may result in an unacceptable health risk.

Critical limit - the maximum or minimum value to which a physical, biological, or chemical parameter must be controlled at a critical control point to minimize the risk that the identified food safety hazard may occur.

Cross contamination - transfer of harmful organisms between items by direct or indirect contact.

Cross connection - any physical link through which contaminants from drains, sewers, or waste pipes can enter a potable water supply.

Cryptosporidium parvum - single-cell parasites found in animal feces and contaminated water.

Cryovac - a proprietary term for vacuum packaging material, which has entered the language to mean all vacuum packaging, like "Kleenex" or "Band-Aid."

Cumulative - increasing in effect by successive additions. For example, hot food must be cooled from 135°F (57°C) to 70°F (21°C) within 2 hours; it must reach 41°F (5°C) within 6 hours to prevent bacterial growth. Therefore, the cumulative time the food is in this danger zone equals a total of 6 hours or less.

Cut leafy greens - fresh leafy greens whose leaves have been cut, shredded, sliced, chopped, or torn. The term "leafy greens" includes iceberg lettuce, romaine lettuce, leaf lettuce, butter lettuce, baby leaf lettuce (i.e., immature lettuce or leafy greens), escarole, endive, spring mix, spinach, cabbage, kale, arugula and chard. The term "leafy greens" does not include herbs such as cilantro or parsley.

Cutting - opening or sampling a product to evaluate its appearance, flavor, quality, and/or consistency.

Cyclospora cayetanensis - a parasite that is found in contaminated water and fresh produce.

Dairy/deli case extender - an insulated container display attached to a refrigerated case that extends into an aisle to stimulate impulse buys.

Danger zone - temperatures between 41°F (5°C) and 135°F (57°C).

Date coding or marking - a date and source code printed on an item to indicate its shelf life. Date codes assist with quality control (first-in, first-out) and proper stock rotation. Date coding may also apply to affixing a "sell by" or "pull by" date on merchandise, which is on display (as in the bakery department) to prevent the sale of off-quality products.

Disinfectant - destroys harmful bacteria.

Dead man's switch - activates equipment when depressed and stops if pressure is relieved.

Deli-bake - a combination in-store bakery and deli department where equipment, floor space, and labor are shared, usually under common supervision.

Delicatessen - an in-store department with cooked foods, salads, cold cuts, and cheeses, etc.

Demo or demonstration - a product promotion in a store with samples to eat and cooking tip handouts or coupons.

Detergents - cleaning agent, which contains surfactants used with water to break down soil to make it easier to remove.

Deviation - to diverge; to go in different directions.

Digital thermometer - a battery powered thermometer that reveals temperature in a digital numerical display.

Disclosure - a written statement that clearly identifies the animal-derived foods which are, or can be ordered, raw, undercooked, or without otherwise being processed to eliminate pathogens, or items that contain an ingredient that is raw, undercooked, or without otherwise being processed to eliminate pathogens.

Dock - an area to receive, load, and unload shipments.

Dressed fish - a whole scaled, cleaned fish, sold with or without the head.

Drinking water - refers to water that is safe to drink. Drinking water is traditionally known as "potable water," and it meets all criteria as specified in *40 CFR 141–National Primary Drinking Water Regulations.*

Dry grocery - nonperishable grocery products.

Dry grocery nonfoods - products that are not food, such as paper products, detergents, or pet items.

Easy-to-clean - materials and design that facilitate cleaning.

Easy-to-move - on wheels, raised on legs, or otherwise designed to facilitate cleaning.

Employee - person working in or for a food establishment who engages in food preparation, service, or other assigned activity.

Enterohemorrhagic *Escherichia coli* (EHEC) - *E. coli* bacteria that produce bloody diarrhea (hemorrhagic colitis) and hemolytic uremic syndrome (HUS). EHEC are a subset of Shiga toxin-producing *E. coli* bacteria. Some examples of serotypes of EHEC include *E. coli* O157:H7 and *E. coli* O157:NM.

Equipment - the appliances: stoves, ovens, etc.; and storage containers, such as refrigerated units used in food establishments.

Ethnic foods - products that a particular ethnic (racial, national, religious, or cultural) group favor, such as Mexican, Chinese, or Kosher.

Evaluation procedures - systematically checking progress to determine if goals have been met.

Exotic produce - fruits and vegetables not grown in North America and considered exotic.

Expiration date - a manufacturer's "sell-by" date stamped on products to indicate shelf life.

Facultative anaerobe - an organism that can grow with or without free oxygen.

FATTOM - an acronym used to indicate the six conditions bacteria need for growth. These conditions are food, acid, temperature, time, oxygen, and moisture.

FDA *Food Code* - list of suggested regulations that may be used by state, local, and tribal jurisdictions when making their own regulations.

FIFO - acronym for first-in, first-out, used to describe stock rotation procedures of using older products first.

Finger cot - waterproof covering for one finger.

FMI (Food Marketing Institute) - an international trade association of independent grocers, chain stores, and wholesalers.

Food additive - Substance added to food that is from the Generally Regarded as Safe (GRAS) list.

Food allergen - a substance in food that causes the human immune system to produce chemicals and histamines in order to protect the body. These chemicals produce allergic symptoms that affect the respiratory system, gastrointestinal tract, skin, or cardiovascular system. See major food allergen.

Food establishment - an operation that stores, prepares, packages, serves, vends, or otherwise provides food for human consumption, such as a restaurant, food market, institutional feeding location, or vending location.

Food recall - the FDA and USDA can request a recall of a food product if the agency has determined a potential hazard exists.

Food safety - use of measures to keep food from becoming contaminated. Such as time temperature control, prevention of cross contamination, etc.

Foodborne disease outbreak - an incident in which two or more people experience a similar illness after ingesting a common food, and in which epidemiological analysis identifies the food as the source of the illness.

Foodborne illness - an illness caused by the consumption of a contaminated food.

Food-contact surface - any surface of equipment or any utensils that food normally touches.

Foot-candle - unit of lighting equal to the illumination one foot from a uniform light source.

Forklift - a vehicle with projecting prongs that slide under a pallet to move merchandise in a warehouse or store.

Fresh - just picked, gathered, produced, live or unprocessed, not stale food. A term associated with perimeter departments, including produce, deli, bakery, or floral; also, unfrozen.

Fungi - a group of microorganisms that includes mold and yeasts.

Giardia lamblia - single-cell parasite found in animal feces and contaminated water and causes foodborne illness.

Garbage - wet waste matter, usually food product, that cannot be recycled.

Garnish - a decoration on salads, such as sprigs of watercress, lettuce, or other colorful items.

Gastroenteritis - an inflammation of the linings of the stomach and intestines, which can cause nausea, vomiting, abdominal cramping, and diarrhea.

General merchandise (GM) - products other than food sold in supermarkets that require special buying, warehousing, and servicing. GM classes are: hardiness, softness, reading/writing lines, health and beauty care (HBC), and services.

Generally recognized as safe (GRAS) - a food safety FDA term that indicates all ingredients are approved for human consumption. (See GRAS substances.)

Germicide - a substance that kills harmful microbes and germs.

Germs - general term for microorganisms, including bacteria and viruses.

Grade - a food industry classification system or standard that indicates a quality level, such as Grade A, Prime, or Extra Fancy.

Grade standards - primarily standards of quality to help producers, wholesalers, retailers, and customers in marketing and purchasing food products. The grade standards are not aimed at protecting the health of the consumer but rather at ensuring value received according to uniform quality standards.

GRAS substances - GRAS stands for generally recognized as safe. These are substances added to foods that have been shown to be safe based on a long history of common usage in food.

Grocery store - a retail store that sells a variety of food products, including some perishable items and general merchandise.

Grocery wholesaler - a middleman who buys food and supplies from manufacturers to resell in smaller quantities to retailers; cooperatives and voluntaries are the two major types.

HACCP - Hazard analysis critical control point.

Hair restraints - Cap, net, or beard cover to contain hair.

Hand antiseptics - Antimicrobial substances that are applied to the skin to reduce the number of microbes. Hand antiseptics may be used as a topical application, a hand antiseptic solution used as a hand dip, or a hand antiseptic soap.

Hand washing - the proper cleaning of hands with soap and warm water to remove dirt, filth, and disease germs.

Handwashing sink - a lavatory or wash basin installed for use in personal hygiene and designed for the washing of the hands. Includes an automatic handwashing facility.

Harborage - shelter for pests.

Hard water - caused by dissolved salts such calcium, magnesium and iron.

Hazard - a biological, chemical, or physical agent that may cause an unacceptable consumer health risk.

Hazard analysis - to identify hazards (problems) that might be introduced into food by unsafe practices or the intended use of the product.

Hazard analysis critical control point (HACCP) - a federal guideline to ensure safe food handling and preparation from receiving to point of sale. See HACCP.

Heat-and-eat - a precooked food that requires heating before consumption.

Hepatitis A virus - a foodborne virus that causes foodborne illness.

Hermetic packaging - a container completely sealed by heat against the entry of bacteria, molds, yeasts, and filth as long as it remains intact.

Highly susceptible population (HSP) - persons who are more likely than other people in the general population to experience foodborne disease because they are immunocompromised, preschool-age children, pregant or lactating women, or older adults and are obtaining food at a facility such as a hospital, nursing home, custodial care, or senior center.

Home meal replacement - foods prepared in a store and consumed at home or in-store, which require little or no preparation on the part of the consumer.

Host - an animal or plant that harbors or nourishes parasites.

Hot-holding - refers to the safe temperature range of 135°F (57°C) and above to maintain properly cooked foods hot until served.

Housekeeping - operational procedures to ensure cleanliness, safety, sanitation, and maintenance for a store or warehouse.

Immunocompromised - the term used to describe individuals, such as infants, children, pregnant women, and those with weakened immune systems, for whom foodborne illness can be very severe, even life threatening.

Impermeable - does not permit passage, especially of fluids.

Individually quick frozen (IQF) - a food processing technique that freezes products in the final stage of processing. It is then wrapped and packaged for shipment.

Infection - illness caused by eating food that contains living disease-causing microorganisms.

Infestation - presence of a large number of pests.

Injected - manipulating meat to which a solution is introduced into the interior by processes referred to as "injecting," "pump marinating," or "stitch pumping."

Inspection for wholesomeness - an examination of domestic and imported meat, poultry, and egg products to assure they are wholesome and free from adulteration.

Integrated pest management (IPM) - a system of preventive and control measures used to control or eliminate pest infestations in food establishments.

Intoxication - illness caused by eating food that contains a harmful chemical or toxin.

Irradiation - a food preservation process that utilizes radiation to control bacteria growth and increase shelf life.

Kick plate (base) - a metal sheet, usually at the bottom of doors, for protection purposes.

Kitchenware - utensils used to prepare food.

Larva - immature stage of development of insects and parasites.

Leftovers - any food that is prepared for a particular meal and is held over for service at a future meal.

Listeria monocytogenes - A facultative microbe that causes foodborne illness.

Low temp - A refrigerator that holds products at a below-freezing temperature [32°F (0°C)] or less.

Major food allergen - milk, egg, fish (such as bass, flounder, cod), crustacean shellfish (such as crab, lobster, or shrimp), tree nuts (such as almonds, pecans, or walnuts), wheat, peanuts, and soybeans; or a food ingredient that contains protein obtained from one of these foods.

Manager - the individual present at a foodservice establishment who supervises employees who are responsible for the storage, preparation, display, and service of food to the public.

MAP - modified atmosphere packaging.

Measuring device - usually a thermometer that registers the temperature of products and water used for sanitizing.

Mechanically tenderized - means manipulating meat with deep penetration by processes which may be referred to as "blade tenderizing," "jaccarding," "pinning," "needling," or using blades, pins, needles or any mechanical device.

Media - newspapers, magazines, radio, and television.

Mesophile - microorganism that grows best at moderate temperatures.

Microbe - a microorganism that can cause disease. Bacteria, molds, and yeast that can grow on various food and equipment surfaces; the main cause of discoloration in meat and food poisoning.

Microorganism - bacteria, viruses, molds, and other tiny organisms that are too small to be seen with the naked eye. The organisms are also referred to as microbes because they cannot be seen without the aid of a microscope.

Misbranding - falsely or misleadingly packaged or labeled food; may contain ingredients not included on the label or does not meet national standards for that food.

Modified atmosphere packaging (MAP) - a food-processing technique where foods are placed in a flexible container and the air is removed from the package. Gases may be added to help preserve the food. See MAP.

Mold - any of various fungi that spoil food and have a fuzzy appearance.

Monitoring procedures - a defined method of checking foods during receiving, storage, preparation, holding, and serving processes.

Non–food–contact surface - any area not designed to touch food.

Norovirus or **Norwalk virus** - common foodborne virus found in contaminated water, produce, seafood, and ready-to-eat foods.

Nutrition labeling - an accurate list of ingredients printed on food, beverage, and drug labels.

Occupational Safety and Health Administration (OSHA) - a federal agency that sets work-place safety standards and inspects facilities for safe working conditions.

Onset time - the period between eating a contaminated food and developing symptoms of a foodborne illness.

Open dating - a date stamped or printed on the label of perishable items to indicate a pull date (a date by which the item must be sold or removed from the shelf) or pack date (the date the item was packaged). Clear, readable dates that are printed on labels.

Organically grown - an imprecise term that means a grower did not use chemicals or that a processor did not use preservatives in the product.

Ovenable - a food ready to be heated, either in an oven or microwave.

Overstock - an excessive amount of product purchased in anticipation of increased sales volume.

Overwrap - to wrap a plastic container in cellophane to prevent tampering.

Pack date - the date on which a product was made or packaged for sale.

Palatable - having an acceptable taste and flavor.

Pallet - a standard-sized base for assembling, sorting, stacking, handling, and transporting goods as a unit. The industry standard is GPCspec- 4-way entry, 48 inches x 40 inches hardwood pallets.

Parasite - an animal or plant that lives in or on another and from whose body it obtains nourishment.

Parts per million (ppm) - unit of measure for water hardness and chemical sanitizing solution concentrations.

Pasteurization - a low-heat treatment used to destroy disease-causing organisms and/or extend the shelf life of a product by destroying organisms and enzymes that cause spoilage.

Pathogenic - capable of causing disease; harmful; any disease-causing agent.

Perishables - foods requiring refrigeration or special handling because they spoil easily, such as meat, seafood, produce, deli, bakery, and dairy.

Person in charge - the individual present at a food establishment who is responsible for the operation at the time of inspection.

Personal hygiene - health habits including bathing, washing hair, wearing clean clothing, and proper hand washing.

Pest control operator - licensed individual or certified technician who provides pest control services.

pH - the symbol that describes the acidity or alkalinity of a substance, such as food.

Physical hazard - particles or fragments of items not supposed to be in foods.

Point of sale (POS) - the place in a retail store where products are scanned through the register system, data is collected, and sales are tendered. POS also describes sales data generated by checkout scanners.

Point of purchase - the locations within a retail store where a customer purchases products.

Potable water - water that is safe to drink. Also known as drinking water.

Potentially hazardous food (time/temperature control for safety food) - a food that requires time/temperature control for safety (TCS) to limit the growth of pathogenic microorganism or toxin formation. Potentially hazardous food (time/temperature control for safety food) [PHF (TCS)] includes an animal food that is raw or heat-treated; a plant food that is heat-treated or consists of raw seed sprouts, cut melons, cut tomatoes, cut leafy greens, or garlic-in-oil mixtures that are not modified in a way that results in mixtures that do not support pathogenic microorganism growth or toxin formation. A food that because of the interaction of its A_w and pH values is designated as Product Assessment required (PA) should be considered a PHF (TCS) until further study proves otherwise.

Pounds per square inch (psi) - amount of pressure per square inch.

Premises - physical environment of the food establishment; grounds, interior, and exterior of building(s).

Preventive measures - procedures that keep foods from becoming contaminated.

Priority foundation item - a provision in the FDA *Food Code* whose application supports, facilitates or enables one or more priority items.

Priority item - a provision in the FDA *Food Code* whose application contributes directly to the elimination, prevention or reduction to an acceptable level, hazards associated with foodborne illness or injury and there is no other provision that more directly controls the hazard.

Proof box - a piece of equipment in which heat and humidity are controlled in order for dough to rise in preparation for baking.

Psychrophiles - microorganisms that grow best at cold temperatures.

Pull date - the date by which a product must be either sold or pulled from a shelf.

Quaternary ammonium - a chemical sanitizing compound that is relatively safe for skin contact and is generally noncorrosive; effective in both acid and alkaline solutions.

Ratite - a flightless bird, such as an emu, ostrich, or rhea.

Reach-in case - a refrigerated display case with a self-service door used for perishable products.

Ready-to-eat foods - products that are in a form that is edible without washing, cooking, or additional preparation by the food establishment or the customer.

Receiver - an authorized associate of a warehouse or retail store who receives and checks deliveries for condition and accurate amount. The first handler of the delivery receipt or invoice.

Receiving - a door or dock of a warehouse or store designated for receiving merchandise from a supplier. The procedure for physically and legally accepting a shipment of product.

Receiving clerk - see receiver.

Receiving log - the record or listing of products received with appropriate entries.

Reclaimed goods - unsalable product at the time of delivery that is returned to a wholesaler/vendor for reclamation.

Reconstitute - to combine dehydrated foods with water or other liquids to bring back to its original state; example, addition of water to powdered milk.

Refuse - trash, rubbish, waste.

Reheating - the act of providing sufficient heat [at least 165°F or (74°C)] within a 2-hour time period to ensure the destruction of any foodborne pathogens that may be present in that cooked and cooled food.

Re-service - the transfer of food that is unused and returned by a consumer after being served or sold and in the possession of the consumer, to another person.

Returns - unsold, damaged, or defective merchandise sent to a supplier or distributor for credit or refund.

Rework - perishables: to crisp or trim a product that looks case-worn, grocery: to reaffix labels. To refine a category or shelf set.

Re-wraps - products that are removed, reconditioned (if salable), and displayed with limited sell-by dates.

Risk - the chance of injury, damage, or loss.

Rotation - a shelf-stocking procedure that assures first-in, first-out by pulling older stock forward and placing newer stock at the back during restocking.

Rotavirus - viral infection associated with sewage, contaminated food, and poor hand washing, and found especially in infants and children.

Rotisserie - a rotating grill with an electrically turned spit that cooks meats.

Safety cutter - a case cutter used to open cases of product.

Salmonella **spp.** - facultative anaerobe that causes foodborne illness.

Salvage - product containers/shippers (bales, pallets, containers) that must be returned or recycled to defray operational costs.

Sanitary - healthful and hygienic. The number of harmful microorganisms and other contaminants have been reduced to safe levels.

Sanitation - maintenance of conditions that are clean and promote good health.

Sanitizer - approved substance or method to use when sanitizing.

Sanitizing - application of an agent that reduces microbes to safe levels.

Scombrotoxin - seafood toxin originating from histamine-producing bacteria.

Sealed - closed tightly.

Segregation - locating general merchandise products (GM) in a well-defined area of a store rather than in aisles next to or across from food products.

Selectivity - chemical sanitizers, especially quats, may kill only certain organisms and not others.

Sensitive ingredient - an ingredient that is prone to support bacterial growth.

Shelf life - time period a product can be expected to maintain maximum quality and freshness.

Shelf stable - a processed food product that remains safe to eat without refrigeration.

Shellfish - an aquatic animal. Molluscan shellfish include clams, oysters, mussels, squid, octopus, and scallops. Crustacean shellfish include crabs, lobster, and shrimp.

Shellstock - raw, in-shell molluscan shellfish.

Shiga toxin-producing *Escherichia coli* - any *E. coli* capable of producing Shiga toxins (also called verocytotoxins or "Shiga-like" toxins). This includes, but is not limited to, *E. coli* reported as serotype O157:H7, O157:NM, and O157:H.

Shigella **spp.** - facultative anaerobic bacteria that cause foodborne illness.

Shucking - a process of opening shellfish and removing the animal from the shell, such as oysters, clams, mussels, etc.

Silicate - salt or ester derived from silica; a hard, glassy material found in sand.

Single-use articles - items intended for one use and then discarded, such as paper cups or plastic eating utensils.

Slacking - the process used to moderate the temperature of a food, such as allowing a food to gradually increase from -10°F (-23°C) to 25°F (-4°C) in preparation for deep-fat frying or to facilitate even heat penetration during the cooking or reheating of a previously block-frozen food, such as spinach.

Sneeze guard - a clear, solid barrier that partially covers food in self-service areas to keep customers from coughing, sneezing, or projecting droplets of saliva directly onto food.

Soil - dirt and filth.

Sous vide - a European food-packaging technique in which a prepared product is placed in individual pouches, cooked under a vacuum, and quickly chilled. Products are frozen or refrigerated until used.

Splash contact surfaces - areas that are easy to clean and designed to catch splashed substances in work areas.

Spoilage - significant food deterioration, usually caused by bacteria and enzymes, that produces a noticeable change in the taste, odor, or appearance of the product.

Spoils - goods that cannot be sold for which a retailer receives a credit from a supplier. Also called stales.

Spore - the inactive or dormant state of some rod-shaped bacteria.

Standard of identity - Food and Drug Administration (FDA) standards for food composition.

Standard operating procedures - a comprehensive book of a company's policies and procedures. Also called SOPs.

Staphlococcus aureus - facultative anaerobic bacterium that produces a heat stable toxin as it grows on food.

Stationary equipment - equipment that is permanently fastened to the floor, table, or counter top.

Sulfites - preservatives used to maintain freshness and color of fresh fruits and vegetables; subject to state regulations.

Supermarket - a conventional grocery store, but not a warehouse club or mass merchant, with annual sales of $2 million or more per store.

Supplier - a generic term for wholesalers, who sell to and supply retailers directly and indirectly (i.e., manufacturer, vendor, broker, reseller).

Surfactant - chemical agent in detergent that reduces the surface tension, allowing the detergent to penetrate and soak soil loose; wetting agent.

Swells - unsalable items with expanded containers or lids signifying faulty food handling, processing, or sealing.

Tableware - plates, cups, bowls, etc.

Task analysis - examining a job task to determine what it takes to do the job.

Temperature abuse - allowing foods to remain in the temperature danger zone [41°F (5°C) to 135°F (57°C)] for an unacceptable period of time.

Temperature danger zone - temperatures between 41°F (5°C) and 135°F (57°C) at which bacteria grow best.

Test kit - device that accurately measures the concentration of sanitizing solutions to assure they are at proper levels.

Thermocouple - a temperature-measuring device that consists of two different types of metals, joined together at one end. When the junction of the two metals is heated or cooled, a voltage is created that can be correlated back to the temperature. Thermocouples are the preferred temperature-measuring devices for use with all foods.

Trichinella spiralis - Foodborne roundworm that causes a foodborne infection.

Unsalables - products unworthy of sale (i.e., damaged, out of date, spoiled).

Upright freezer - an upright refrigerated display unit with doors used for merchandising frozen foods.

USDA Grade - U.S. Department of Agriculture grades that relate to a specified quality of product. Grade denotes quality and USDA denotes product inspected for wholesomeness.

U.S. grade stamp - signifies that a product is clean, safe, and wholesome, and has been produced in an acceptable establishment, with the appropriate equipment, under the supervision of federal inspectors. It also indicates the product is of a specific grade.

U.S. Department of Agriculture (USDA) - a federal agency that oversees food production and inspection. The USDA establishes grade standards for commodities, conducts agricultural research and makes results available, administers food programs (i.e., food stamps), and distributes food.

U.S. Department of Commerce (USDC) - a federal agency that oversees trade and competition. The USDC establishes grade standards for seafood commodities, conducts agricultural research, and makes results available.

Utensil - a food-contact implement or container used in the storage, preparation, transportation, dispensing, sale, or service of food, such as kitchenware or tableware that is multiuse, single-service, or single-use; gloves used in contact with food; temperature-sensing probes of food temperature-measuring devices; and probe-type price or identification tags used in contact with food.

Vacuum breaker - designed for use under a continuous supply of pressure. Spring-loaded device to operate after extended periods of hydrostatic pressure.

Vacuum packaging - a packaging process in which air is removed from a package as it is sealed.

Vegetative state - the active state of a bacterium in which the cell takes in nourishment, grows, and produces wastes.

Ventilation - air circulation that removes smoke, odors, moisture, and grease-laden vapors from a room and replaces them with fresh air.

Verification - to prove to be true by evidence, usually a record of times and temperatures of food from receiving to serving or vending.

Vibrio ssp. - group of three organisms (*Vibrio* cholera, *Vibrio* parahaemolyticus, *Vibrio* vulnificus) found in raw or improperly cooked fish and shellfish that can cause foodborne illness.

Viruses - any of a group of infectious microorganisms that reproduce only in living cells. They cause diseases, such as mumps and Hepatitis A virus, and can be transmitted through food.

Warewashing - the cleaning and sanitizing of utensils and food-contact surfaces of equipment.

Water activity (A_w) - a measure of the free moisture in a food. Pure water has a water activity of 1.0, and potentially hazardous foods have a water activity of 0.85 and higher.

Waxing - applying an edible wax to some fruits and vegetables to help maintain a fresh, bright appearance and to preserve product quality.

Wetting agent - a substance that breaks down the soil to allow water and soap or detergent to liquefy and remove dirt and grease.

Whole-muscle, intact beef - whole-muscle beef that is not injected, mechanically tenderized, reconstructed, or scored and marinated, from which beef steaks may be cut.

Wholesome - something that is favorable to or promotes health.

Yeast - type of fungus that is not known to cause illness when present in foods but can cause damage to food products and will change taste; useful in making products, such as bread and beer.

Appendix A

Answers to End of Chapter Questions

Chapter 1 Answers

1. b	6. d
2. d	7. d
3. b	
4. d	
5. c	

Chapter 2 Answers

1. b	6. b
2. a	7. c
3. d	8. a
4. c	9. b
5. c	10. b

Chapter 3 Answers

1. d	7. d
2. b	8. d
3. a	9. c
4. a	10. d
5. b	11. c
6. b	12. a

Chapter 4 Answers

1. c	9. c
2. b	10. d
3. c	11. c
4. a	12. d
5. a	13. d
6. b	14. c
7. d	15. b
8. c	

Chapter 5 Answers

1. d	6. b
2. b	7. c
3. a	8. d
4. b	9. a
5. d	10. d

Chapter 6 Answers

1. d	6. d
2. c	7. a
3. b	8. b
4. a	9. a
5. c	10. d

Chapter 7 Answers

1. a	6. b
2. d	7. d
3. d	8. a
4. a	9. b
5. c	10. d

Chapter 8 Answers

1. a	6. a
2. b	7. b
3. d	8. d
4. b	9. d
5. b	

Chapter 9 Answers

1. a	6. c
2. d	7. b
3. c	8. d
4. b	9. c
5. d	10. d

Appendix B

The Process for Determining Potentially Hazardous Food (Time/Temperature Control for Safety Food [PHF (TCS)])

The information appearing below was taken from pages 320 and 321 of "Annex 3—Public Health Reasons/Administrative Guidelines" of the 2009 *Food Code*. Additional information about PHF (TCS) and how to use the decision tree and tables presented in Annex 3 of the *Food Code* to determine if an item is a PHF (TCS) can be found at the following web site: www.fda.gov/Food/FoodSafety/RetailFoodProtection/FoodCode/FoodCode2009/ucm189170.htm

Instructions for using the decision tree and Table A and Table B:

1. Does the operator want to hold the food without using time or temperature control?

   a. No—Continue holding the food at ≤ 41°F (5°C) or ≥135°F (57°C) for safety and/or quality.

   b. Yes—Continue using the decision tree to identify which table to use to determine whether time/temperature control for safety (TCS) is required.

2. Is the food heat-treated?

   a. No—The food is either raw, partially cooked (not cooked to the temperature specified in section 3-401.11 of the *Food Code*) or treated with some method other than heat. Proceed to step 3.

   b. Yes—If the food is heat-treated to the required temperature for that food as specified under section 3-401.11 of the *Food Code*, vegetative cells will be destroyed although spores will survive. Proceed to step 4.

3. Is the food treated using some other method?

   a. No —The food is raw or has only received a partial cook allowing vegetative cells and spores to survive. Proceed to step 6.

   b. Yes—If a method other than heat is used to destroy pathogens, such as irradiation, high-pressure processing, pulsed light, ultrasound, inductive heating, or ozonation, the effectiveness of the process needs to be validated by inoculation studies or other means. Proceed to step 7.

4. Is the food packaged to prevent recontamination?

   a. No—Recontamination of the product can occur after heat treatment because it is not packaged. Proceed to step 6.

   b. Yes—If the food is packaged immediately after heat treatment to prevent recontamination, higher ranges of pH and/or A_w can be tolerated because sporeforming bacteria are the only microbial hazard. Proceed to step 7.

5. Further product assessment or vendor documentation required.

    a. The vendor of this product may be able to supply documentation that inoculation studies indicate the food can be safely held without TCS.

    b. Food prepared or processed using new technologies may be held without time/temperature control provided the effectiveness of the use of such technologies is based on a validated inoculation study.

6. Using the food's known pH and/or A_w values, position the food in the appropriate table.

    a. Choose the column under "pH values" that contains the pH value of the food in question.

    b. Choose the row under "A_w values" that contains the A_w value of the food in question.

    c. Note where the row and column intersect to identify whether the food is "non-PHF/non-TCS food" and therefore does not require time/temperature control, or whether further product assessment (PA) is required. Other factors, such as redox potential, competitive microorganisms, salt content, or processing methods may allow the product to be held without time/temperature control, but an inoculation study is required.

7. Use **Table A** for foods that are heat-treated and packaged **OR** use **Table B** for foods that are not heat-treated or heat-treated but not packaged.

8. Determine if the item is non-PHF/non-TCS or needs further product assessment (PA).

Table A Interaction of pH and A_w for control of spores in food heat-treated to destroy vegetative cells and subsequently packaged

A_w values	pH values		
	4.6 or less	>4.6 – 5.6	>5.6
≤0.92	non-PHF*/non-TCS food**	non-PHF/non-TCS food	non-PHF/non-TCS food
> 0.92 – 0.95	non-PHF/non-TCS food	non-PHF/non-TCS food	PA***
> 0.95	non-PHF/non-TCS food	PA	PA

* PHF means Potentially Hazardous Food
** TCS food means Time/Temperature Control for Safety Food
*** PA means Product Assessment required

Table B Interaction of pH and A_w for control of vegetative cells and spores in food not heat-treated or heat-treated but not packaged

A_w values	pH values			
	< 4.2	4.2 – 4.6	> 4.6 – 5.0	> 5.0
< 0.88	non-PHF*/ non-TCS food**	non-PHF/non-TCS food	non-PHF/non-TCS food	non-PHF/non-TCS food
0.88 – 0.90	non-PHF/non-TCS food	non-PHF/non-TCS food	non-PHF/non-TCS food	PA***
>0.90 – 0.92	non-PHF/non-TCS food	non-PHF/non-TCS food	PA	PA
>0.92	non-PHF/non-TCS food	PA	PA	PA

* PHF means Potentially Hazardous Food
** TCS food means Time/Temperature Control for Safety Food
*** PA means Product Assessment required

Before using Tables A and B in determining whether a food requires TCS, answers to the following questions should be considered:

◆ Is the intent to hold the food without using time or temperature control?

◆ If the answer is no, no further action is required. The decision tree presented at the beginning of the Appendix is not needed to determine if the item is a PHF (TCS).

◆ Is the food raw or is the food heat-treated?

◆ Does the food already require TCS according to the definition of PHF (TCS)?

◆ Does a product history with sound scientific rationale exist indicating a safe history of use?

◆ Is the food processed and packaged so that it no longer requires TCS, such as ultrahigh temperature (UHT) creamers or shelf-stable canned goods?

◆ What is the pH and A_w of the food in question using an independent laboratory and Association of Official Analytical Chemists (AOAC) methods of analysis?

A food designated as meeting PA, in either Table A or B should be considered PHF until further study proves otherwise. The PA means that based on the food's pH and A_w and whether it was raw or heat-treated or packaged, it has to be considered PHF until inoculation studies or some other acceptable evidence shows that the food is a PHF (TCS) or not. The *Food Code* requires a variance request to be sent in advance to the regulatory authority with the evidence that the food does not require TCS.

Appendix C

Summary of Agents That Cause Foodborne Illness

Agents that Cause Foodborne Illness				
CAUSATIVE AGENT (*Sporeforming bacteria)	**TYPE OF ILLNESS**	**SYMPTOMS ONSET**	**COMMON FOODS**	**PREVENTION**
Anisakis spp.	Parasitic infection	Coughing, vomiting 1 hr to 2 weeks	Raw or undercooked seafood, especially bottom-feeding fish	Cook fish to the proper temperature throughout; freeze to meet 2009 FDA *Food Code* specifications.
Bacillus cereus	Bacterial intoxication or toxin-mediated infection	(1) Diarrhea, abdominal cramps (8 to 16 hrs) (2) Vomiting type: vomiting, diarrhea, abdominal cramps (30 minutes to 6 hrs)	(1) Diarrhea type: meats, milk, vegetables (2) Vomiting type: rice, starchy foods, grains and cereals	Properly heat, cool, and reheat foods.
Campylobacter jejuni	Bacterial infection	Watery, bloody diarrhea (2 to 5 days)	Raw chicken, raw milk, raw meat	Properly handle and cook foods; avoid cross contamination.
Ciguatoxin	Fish toxin, originating from toxic algae of tropical waters	Vertigo, hot/cold flashes, diarrhea, vomiting (15 minutes to 24 hrs)	Marine finfish including grouper, barracuda, snappers, jacks, mackerel, triggerfish, reef fish	Purchase fish from a reputable supplier; cooking WILL NOT inactivate the toxin.
Clostridium botulinum	Bacterial intoxication	Dizziness, double vision, difficulty in breathing and swallowing, headache (12 to 36 hrs)	Improperly canned foods, vacuum-packed refrigerated foods, cooked foods in anaerobic mass	Properly heat process anaerobically packed foods; DO NOT use home canned foods.
Clostridium perfringens	Bacterial toxin-mediated infection	Intense abdominal pains and severe diarrhea (8 to 22 hrs)	Spices, gravy, improperly cooled foods (especially meats and gravy dishes)	Properly cook, cool, and reheat foods.
Cryptosporidium parvum	Parasitic infection	Severe watery diarrhea (within 1 week of ingestion)	Contaminated water, food contaminated by infected food employees	Use potable water supply; practice good personal hygiene and hand washing.
Cyclospora cayetanensis	Parasitic infection	Watery and explosive diarrhea, loss of appetite, bloating (1 week)	Water, strawberries, raspberries, and raw vegetables	Good sanitation; reputable supplier.
Food allergens	An allergic reaction usually involving the skin, mouth, digestive tract, or airways	Skin: hives, rashes, and itching Mouth: swelling and itching of the lips and tongue Digestive tract: vomiting and diarrhea Airways: difficulty breathing, wheezing	Foods that contain: milk, egg, wheat, tree nuts and peanuts, soybeans, fish, and shellfish	Packaged and prepared foods must be properly labeled if they contain common food allergens so sensitive people can avoid them.
Giardia lamblia	Parasitic infection	Diarrhea within 1 week of contact	Contaminated water	Potable water supply; good personal hygiene and hand washing (cont.)

Agents that Cause Foodborne Illness				(continued)
CAUSATIVE AGENT (*Sporeforming bacteria)	**TYPE OF ILLNESS**	**SYMPTOMS ONSET**	**COMMON FOODS**	**PREVENTION**
Hepatitis A virus	Viral infection	Fever, nausea, vomiting, abdominal pain, fatigue, swelling of the liver, jaundice (15 to 50 days)	Foods that are prepared with human contact, contaminated water	Wash hands and practice good personal hygiene; avoid raw seafood.
Listeria monocytogenes	Bacterial infection	(1) Healthy adult: flu-like symptoms (2) Highly susceptible population: septicemia, meningitis, encephalitis, birth defects (1 day to 3 weeks)	Raw milk, dairy items, raw meats, refrigerated ready-to-eat foods, processed ready-to-eat meats, such as hot dogs, raw vegetables, and seafood	Properly store and cook foods; avoid cross contamination; rotate processed refrigerated foods using FIFO to assure timely use.
Mycotoxins	Intoxication	(1)Acute onset: hemorrhage, fluid buildup, possible death (2) Chronic: cancer from small does over time	Moldy grains: corn, corn products, peanuts, pecans, walnuts, and milk	Purchase food from a reputable supplier; keep grains and nuts dry; protect products from humidity.
Norovirus	Viral infection	Vomiting, diarrhea, abdominal pain, headache, and low-grade fever (24 to 48 hrs)	Sewage, contaminated water, contaminated salad ingredients, raw clams, oysters	Use potable water; cook all shellfish; handle food properly; meet time temperature guidelines for PHF (TCS).
Rotavirus	Viral infection	Diarrhea (especially in infants and children), vomiting, low-grade fever, (1 to 3 days (lasts 4 to 8 days)	Sewage, contaminated water, contaminated salad ingredients, raw seafood	Good personal hygiene and hand washing; proper food-handling practices.
Salmonella spp.	Bacterial infection	Nausea, fever, vomiting, abdominal cramps, diarrhea (6 to 48 hrs)	Raw meats, raw poultry, eggs, milk, dairy products	Properly cook foods; avoid cross contamination.
Scombrotoxin	Seafood toxin originating from histamine-producing bacteria	Dizziness, burning feeling in the mouth, facial rash or hives, peppery taste in mouth, headache, itching, and teary eyes, runny nose (1 to 30 minutes)	Tuna, mahi-mahi, bluefish, sardines, mackerel, anchovies, amberjack, abalone	Purchase fish from a reputable supplier; store fish at low temperatures to prevent growth of histamine- producing bacteria; toxin IS NOT inactivated by cooking.
Shellfish toxins: PSP, DSP, DAP, NSP	Intoxication	Numbness of lips, tongue, arms, legs, neck; lack of muscle coordination (10 to 60 minutes)	Contaminated mussels, clams, oysters, scallops	Purchase from a reputable supplier.
Shiga toxin-producing *Escherichia coli*	Bacterial infection or toxin-mediated infection	Bloody diarrhea followed by kidney failure and hemolytic uremic syndrome (HUS) in severe cases (12 to 72 hrs)	Undercooked hamburger, raw milk, unpasteurized apple cider, lettuce, and spinach	Practice good food sanitation; hand washing; properly handle and cook foods.
Shigella spp.	Bacterial infection	Bacillary dysentery, diarrhea, fever, abdominal cramps, dehydration (1 to 7 days)	Foods that are prepared with human contact: salads, raw vegetables, milk, dairy products, raw poultry, non-potable water, ready-to-eat meat	Wash hands and practice good personal hygiene; properly cook foods. (cont.)

Agents that Cause Foodborne Illness				(continued)
CAUSATIVE AGENT (*Sporeforming bacteria)	**TYPE OF ILLNESS**	**SYMPTOMS ONSET**	**COMMON FOODS**	**PREVENTION**
Staphylococcus aureus	Bacterial intoxication	Nausea, vomiting, abdominal cramps, headaches (2 to 6 hrs)	Foods that are prepared with human contact, cooked or processed foods	Wash hands and practice good personal hygiene; cooking WILL NOT inactivate the toxin.
Toxoplasma gondii	Parasitic infection	Mild cases: swollen lymph glands, fever, headache, and muscle aches. Severe cases: may result in damage to the eye or the brain. (10 to 13 days)	Raw meats, raw vegetables, and fruit	Good sanitation; purchase from a reputable supplier; proper cooking.
Trichinella spiralis	Parasitic infection from a nematode worm	Nausea, vomiting, diarrhea, sweating, muscle soreness (2 to 28 days)	Primarily undercooked wild game meats (bear, walrus) and pork products	Cook foods to the proper temperature throughout.
Vibrio spp.	Bacterial infection	Headache, fever, chills, diarrhea, vomiting, severe electrolyte loss, gastroenteritis (2 to 48 hrs)	Raw or improperly cooked fish and shellfish	Practice good sanitation, properly cook foods, avoid serving raw seafood.

Appendix D

Multitiered, Risk-based Employee Health System

The *Food Code* recommends that food establishments implement a risk-based employee health system that will reduce the likelihood that the "Big Five" pathogens—Norovirus, *Salmonella* Typhi, Enterohemorrhagic or Shiga toxin-producing *Escherichia coli*, Hepatitis A virus, and *Shigella* spp.—will be transmitted from infected food employees into food. These agents are known to be readily transmissible via food that has been contaminated by ill food employees and, therefore, are the primary focus of the "Employee Health" section of the *Food Code*.

The *Food Code* recommends a multitiered, risk-based system that balances a food employee's need to work and earn a living with an acceptable health risk for the public. It assures removal of infected food employees when they are most likely to transmit a pathogen to food items and provides guidance on when ill food employees can safely return to work.

Four levels of illness or potential illness have been identified by the FDA as follows:

◆ **Level 1 involves food employees who have specific symptoms (i.e., vomiting, diarrhea, jaundice) while in the work place.** These symptoms are known to be commonly associated with the agents most likely to be transmitted from infected food employees through contamination of food. The first level also relates to employees who have been diagnosed with typhoid fever or an infection with Hepatitis A virus (within 14 days of symptoms). The first level poses the highest potential risk to public health. **The most significant degree of restriction and exclusion applies to the first level of food employee illness.** Infected food employees in Level 1 are likely to be excreting high levels of an infectious pathogen, increasing the chance of transmission to food products and placing the public at higher risk.

◆ **Level 2 relates to employees who have been diagnosed with a specific agent of concern, but who are not currently exhibiting symptoms of disease.** Food employees in Level 2 are still likely to be carrying the infectious agent, but they are less likely to spread the agent into food. However, these employees diagnosed with one of the agents of concern still pose an elevated threat to public health. For this reason, there are a series of exclusions if the employee works in a facility serving highly susceptible populations (HSP) and restrictions (for non-HSP facilities) may be greater, depending on the agent involved.

◆ **Level 3 relates to employees who are diagnosed with a specific agent but never develop any gastrointestinal symptoms.** These food employees are typically identified during a foodborne illness outbreak investigation through microbiological testing. The *Food Code* provides restriction or exclusion guidelines of infected employees who are identified through microbiological testing as carrying a listed pathogen, but are otherwise asymptomatic and clinically well. The exclusion or restriction guidelines are applied until the identified food employees no longer present a risk for foodborne pathogen transmission.

◆ **Level 4 relates to those individuals who are clinically well but who may have been exposed to a listed pathogen and are within the normal incubation period for the disease.** For example, a food employee may have attended a function at which he/she ate food that was associated with an outbreak of shigellosis, but the employee remains well. Such individuals present a lower risk to public health than someone who is either symptomatic or who has a definitive diagnosis. However, they present a level of risk to public health that is greater than if they had not had the exposure. The recommendation in the *Food Code* is to restrict food employees who have had a potential exposure based on the incubation times (time between exposure and the onset of symptoms) of the various agents. As a further protection to public health, it is recommended that such exposed food employees pay particular attention to personal hygiene and report the onset of any symptoms.

This tiered approach links the degree of exclusion and restriction to the degree of risk that an infected food employee will transmit an agent of concern through food. It seeks to strike a balance between protecting public health and meeting the needs of the food employee and employer.

Below is the specific information from the *Food Code* regarding reportable diseases and activities; exclusions and restrictions; and removal, adjustment, or retention of exclusions and restrictions.

2-2 Employee Health

Subpart

2-201 Responsibility of Permit Holder, Person in Charge, Food Employees, and Conditional Employees.

Responsibilities and Reporting Symptoms and Diagnosis	2-201.11 Responsibility of Permit Holder, Person in Charge, and Conditional Employees
	(A) The permit holder shall require food employees and conditional employees to report to the person in charge information about their health and activities as they relate to diseases that are transmissible through food. A food employee or conditional employee shall report the information in a manner that allows the person in charge to reduce the risk of foodborne disease transmission, including providing necessary additional information, such as the date of onset of symptoms and illness, or of a diagnosis without symptoms, if the food employee or conditional employee:

Reportable *symptoms*	(1) Has any of the following symptoms:
	
	(a) Vomiting,
	(b) Diarrhea,
	(c) Jaundice,
	(d) Sore throat with fever, or;
	(e) A lesion containing pus, such as a boil or infected wound that is open or draining and is:
	
	   (i) On the hands or wrists, *unless an impermeable cover, such as a finger cot or stall protects the lesion and a single-use glove is worn over the impermeable cover,*
	   (ii) On exposed portions of the arms, *unless the lesion is protected by an impermeable cover,* or;
	   (iii) On other parts of the body, *unless the lesion is covered by a dry, durable, tight-fitting bandage.*

Reportable *diagnosis*	(2) Has an illness diagnosed by a health practitioner due to: (a) Norovirus, (b) Hepatitis A virus, (c) *Shigella* spp. (d) Enterohemorrhagic or Shiga toxin-producing *Escherichia coli*, or; (e) *Salmonella* Typhi.

Responsibilities and Reporting Symptoms and Diagnosis	2-201.11 Responsibility of Permit Holder, Person in Charge, and Conditional Employees (continued)
Reportable *past illness*	(3) Had a previous illness, diagnosed by a health practitioner, within the past 3 months due to *Salmonella* Typhi, without having received antibiotic therapy, as determined by a health practitioner.
Reportable *history of exposure*	(4) Has been exposed to, or is the suspected source of, a confirmed disease outbreak, because the food employee or conditional employee consumed or prepared food implicated in the outbreak, or consumed food at an event prepared by a person who is infected or ill with: (a) Norovirus within the past 48 hours of the last exposure, (b) Enterohemorrhagic or Shiga toxin-producing *Escherichia coli*, or *Shigella* spp. within the past 3 days of the last exposure, (c) *Salmonella* Typhi within the past 14 days of the last exposure, or; (d) Hepatitis A virus within the past 30 days of the last exposure. (5) Has been exposed by attending or working in a setting where there is a history of confirmed disease outbreak, or living in the same household as, and has knowledge about, an individual who works or attends a setting where there is a confirmed disease outbreak, or living in the same household as, and has knowledge about, an individual diagnosed with an illness caused by: (a) Norovirus within the past 48 hours of the last exposure, (b) Enterohemorrhagic or Shiga toxin-producing *Escherichia coli*, or *Shigella* spp. within the past 3 days of the last exposure, (c) *Salmonella* Typhi within the past 14 days of the last exposure, or; (d) Hepatitis A virus within the past 30 days of the last exposure.
Responsibility of person in charge to *notify the regulatory authority*	(B) The person in charge shall notify the regulatory authority when a food employee is: (1) Jaundiced, or; (2) Diagnosed with an illness due to a pathogen as specified under Subparagraphs (A)(2)(a)—(e) of this section.
Responsibility of the person in charge to *prohibit a conditional employee from becoming a food employee*	(C) The person in charge shall assure that a conditional employee: (1) Who exhibits or reports a symptom, or who reports a diagnosed illness as specified under subparagraphs (A)(1)–(3) of this section, is prohibited from becoming a food employee until the conditional employee meets the criteria for the specific symptoms or diagnosed illness as specified under § 2-201.13, and; (2) Who will work as a food employee in a food establishment that serves a highly susceptible population and reports a history of exposure as specified under subparagraphs (A)(4)–(5), is prohibited from becoming a food employee until the conditional employee meets the criteria as specified under ¶ 2-201.13(I).
Responsibility of the person in charge to *exclude or restrict*	(D) The person in charge shall assure that a food employee who exhibits or reports a symptom, or who reports a diagnosed illness or a history of exposure as specified under subparagraphs (A)(1)–(5) of this section is: (1) Excluded as specified under ¶¶ 2-201.12 (A)–(C), and subparagraphs (D)(1), (E)(1), (F)(1), or (G)(1) and in compliance with the provisions specified under ¶¶ 2-201.13(A)–(G), or; (2) Restricted as specified under subparagraphs 2-201.12(D)(2), (E)(2), (F)(2), (G)(2), or ¶¶ 2-201.12(H) or (I) and in compliance with the provisions specified under ¶¶ 2-201.13(D)–(I).

Responsibilities and Reporting Symptoms and Diagnosis	2-201.11 Responsibility of Permit Holder, Person in Charge, and Conditional Employees (continued)
Responsibility of food employees and conditional employees to *report*	(E) A food employee or conditional employee shall report to the person in charge of the information as specified under ¶ (A) of this section.
Responsibility of food employees to *comply*	(F) A food employee shall: (1) Comply with an exclusion as specified under ¶¶ 2-201.12(A)-(C) and subparagraphs 2-201.12(D)(1), (E)(1), (F)(1), or (G)(1) and with the provisions specified under ¶¶ 2-201.13(A)-(G); or (2) Comply with a restriction as specified under subparagraphs 2-201.12(D)(2), (E)(2), (F)(2), (G)(2), or ¶¶ 2-201.12(H) or (I) and comply with the provisions specified under ¶¶ 2-201.13(D)-(I).

Conditions of Exclusion and Restriction	2-201.12 Exclusions and Restrictions
	The person in charge shall exclude or restrict a food employee from a food establishment in accordance with the following:
Symptomatic with **vomiting or diarrhea**	(A) *Except when the symptom is from a noninfectious condition,* exclude a food employee if the food employee is: (1) Symptomatic with vomiting or diarrhea, or; (2) Symptomatic with vomiting or diarrhea and diagnosed with an infection from norovirus, *Shigella* spp., or Enterohemorrhagic or Shiga toxin-producing *E. coli*.
Jaundiced or diagnosed with **Hepatitis A virus infection**	(B) Exclude a food employee who is: (1) Jaundiced and the onset of jaundice occurred within the last 7 calendar days, *unless the food employee provides to the person in charge written medical documentation from a health practitioner specifying that the jaundice is not caused by Hepatitis A virus or other fecal-orally transmitted infection,* (2) Diagnosed with an infection from Hepatitis A virus within 14 calendar days from the onset of any illness symptoms, or within 7 calendar days of the onset of jaundice, or; (3) Diagnosed with an infection from Hepatitis A virus without developing symptoms.
Diagnosed or reported previous infection due to *S. Typhi*	(C) Exclude a food employee who is diagnosed with an infection from *Salmonella* Typhi, or reports a previous infection with *Salmonella* Typhi within the past 3 months as specified under subparagraph 2-201.11(A)(3).
Diagnosed with an asymptomatic infection from **norovirus**	(D) If a food employee is diagnosed with an infection from norovirus and is asymptomatic: (1) Exclude the food employee who works in a food establishment serving a highly susceptible population, or; (2) Restrict the food employee who works in a food establishment not serving a highly susceptible population.
Diagnosed with *Shigella* **spp.** infection and asymptomatic	(E) If a food employee is diagnosed with an infection from *Shigella* spp. and is asymptomatic: (1) Exclude the food employee who works in a food establishment serving a highly susceptible population, or; (2) Restrict the food employee who works in a food establishment not serving a highly susceptible population.

Conditions of Exclusion and Restriction	2-201.12 Exclusions and Restrictions	(continued)
Diagnosed with **EHEC or STEC** and asymptomatic	(F) If a food employee is diagnosed with an infection from Enterohemorrhagic or Shiga toxin-producing *E. coli*, and is asymptomatic: (1) Exclude the food employee who works in a food establishment serving a highly susceptible population, or; (2) Restrict the food employee who works in a food establishment not serving a highly susceptible population.	
Symptomatic with **sore throat with fever**	(G) If a food employee is ill with symptoms of acute onset of sore throat with fever: (1) Exclude the food employee who works in a food establishment serving a highly susceptible population, or; (2) Restrict the food employee who works in a food establishment not serving a highly susceptible population.	
Symptomatic with **uncovered infected wound or pustular boil**	(H) If a food employee is infected with a skin lesion containing pus, such as a boil or infected wound, that is open or draining and not properly covered as specified under subparagraph 2-201.11(A)(1)(e), restrict the food employee.	
Exposed to foodborne pathogen and works in food establishment serving HSP	(I) If a food employee is exposed to a foodborne pathogen as specified under subparagraphs 2-201.11(A)(4) or (5), restrict the food employee who works in a food establishment serving a highly susceptible population.	

Managing Exclusions and Restrictions	2-201.13 Removal, Adjustment, or Retention of Exclusions and Restrictions
	The person in charge may remove, adjust, or retain the exclusion or restriction of a food employee according to the following conditions: (A) *Except when a food employee is diagnosed with an infection from* Hepatitis A virus *or* Salmonella Typhi:
Removing exclusion for food employee who was symptomatic and not diagnosed	(1) Reinstate a food employee who was excluded as specified under subparagraph 2-201.12(A)(1) if the food employee: (a) Is asymptomatic for at least 24 hours, or; (b) Provides to the person in charge written medical documentation from a health practitioner that states the symptom is from a noninfectious condition.
Norovirus diagnosis	(2) If a food employee was diagnosed with an infection from Norovirus and excluded as specified under subparagraph 2-201.12(A)(2):
Adjusting exclusion for food employee who was symptomatic and is now asymptomatic	(a) Restrict the food employee, who is asymptomatic for at least 24 hours and works in a food establishment not serving a highly susceptible population, until the conditions for reinstatement as specified under subparagraphs (D)(1) or (2) of this section are met, or;
Retaining exclusion for food employee who was symptomatic and is now asymptomatic and works in food establishment serving HSP	(b) Retain the exclusion for the food employee, who is asymptomatic for at least 24 hours and works in a food establishment that serves a highly susceptible population, until the conditions for reinstatement as specified under subparagraphs (D)(1) or (2) of this section are met.
Shigella **spp. diagnosis**	(3) If a food employee was diagnosed with an infection from *Shigella* spp. and excluded as specified under subparagraph 2-201.12(A)(2):
Adjusting exclusion for food employee who was symptomatic and is now asymptomatic	(a) Restrict the food employee, who is asymptomatic for at least 24 hours and works in a food establishment not serving a highly susceptible population, until the conditions for reinstatement as specified under subparagraphs (E)(1) or (2) of this section are met, or;

Managing Exclusions and Restrictions	2-201.13 Removal, Adjustment, or Retention of Exclusions and Restrictions (continued)
Retaining exclusion for food employee who was symptomatic and is now asymptomatic	(b) Retain the exclusion for the food employee, who is asymptomatic for at least 24 hours and works in a food establishment that serves a highly susceptible population, until the conditions for reinstatement as specified under subparagraphs (E)(1) or (2), diagnosis or (E)(1), and (3)(a) of this section are met.
EHEC or STEC diagnosis	(4) If a food employee was diagnosed with an infection from Enterohemorrhagic or Shiga toxin-producing *Escherichia coli* and excluded as specified under subparagraph 2-201.12(A)(2):
Adjusting exclusion for food employee who was symptomatic and is now asymptomatic	(a) Restrict the food employee, who is asymptomatic for at least 24 hours and works in a food establishment not serving a highly susceptible population, until the conditions for reinstatement as specified under subparagraphs (F)(1) or (2) of this section are met, or;
Retaining exclusion for food employee who was symptomatic and is now asymptomatic and works in food establishment serving HSP	(b) Retain the exclusion for the food employee, who is asymptomatic for at least 24 hours and works in a food establishment that serves a highly susceptible population, until the conditions for reinstatement as specified under subparagraphs (F)(1) or (2) are met.
Hepatitis A virus or jaundice diagnosis— removing exclusions	(B) Reinstate a food employee who was excluded as specified under ¶ 2-201.12(B) if the person in charge obtains approval from the regulatory authority and one of the following conditions is met; (1) The food employee has been jaundiced for more than 7 calendar days, (2) The anicteric food employee has been symptomatic with symptoms other than jaundice for more than 14 calendar days, or; (3) The food employee provides to the person in charge written medical documentation from a health practitioner stating the food employee is free of a Hepatitis A virus infection.
S. Typhi diagnosis— **removing exclusions**	(C) Reinstate a food employee who was excluded as specified under ¶ 2-201.12(C) if: (1) The person in charge obtains approval from the regulatory authority, and; (2) The food employee provides to the person in charge written medical documentation from a health practitioner that states the food employee is free from *S. Typhi* infection.

Managing Exclusions and Restrictions	2-201.13 Removal, Adjustment, or Retention of Exclusions and Restrictions (continued)
Norovirus diagnosis—removing exclusion or restriction	(D) Reinstate a food employee who was excluded as specified under subparagraphs 2-201.12(A)(2) or (D)(1) who was RESTRICTED under subparagraph 2-201.12(D)(2) if the person in charge obtains approval from the regulatory authority and one of the following conditions is met: (1) The excluded or restricted food employee provides to the person in charge written medical documentation from a health practitioner stating that the food employee is free of a Norovirus infection, (2) The food employee was excluded or restricted after symptoms of vomiting or diarrhea resolved, and more than 48 hours have passed since the food employee became asymptomatic, or; (3) The food employee was excluded or restricted and did not develop symptoms and more than 48 hours have passed since the food employee was diagnosed.
***Shigella* spp. diagnosis—removing exclusion or restriction**	(E) Reinstate a food employee who was excluded as specified under subparagraphs 2-201.12(A)(2) or (E)(1) or who was restricted under subparagraph 2-201.12(E)(2) if the person in charge obtains approval from the regulatory authority and one of the following conditions is met: (1) The excluded or restricted food employee provides to the person in charge written medical documentation from a health practitioner stating that the food employee is free of a *Shigella* spp. infection based on test results showing two consecutive negative stool specimen cultures that are taken: (a) Not earlier than 48 hours after discontinuance of antibiotics, and; (b) At least 24 hours apart. (2) The food employee was excluded or restricted after symptoms of vomiting or diarrhea resolved, and more than 7 calendar days have passed since the food employee became asymptomatic, or; (3) The food employee was excluded or restricted and did not develop symptoms and more than 7 calendar days have passed since the food employee was diagnosed.
EHEC or STEC diagnosis— removing exclusion or restriction	(F) Reinstate a food employee who was excluded or restricted as specified under subparagraphs 2-201.12(A)(2) or (F)(1) or who was restricted under subparagraph 2-201.12(F)(2) if the person in charge obtains approval from the regulatory authority and one of the following conditions is met: (1) The excluded or restricted food employee provides to the person in charge written medical documentation from a health practitioner stating that the food employee is free of an infection from Enterohemorrhagic or Shiga toxin-producing *Escherichia coli* based on test results that show two consecutive negative stool specimen cultures that are taken: (a) Not earlier than 48 hours after discontinuance of antibiotics; and (b) At least 24 hours apart, (2) The food employee was excluded or restricted after symptoms of vomiting or diarrhea resolved and more than 7 calendar days have passed since the food employee became asymptomatic, or; (3) The food employee was excluded or restricted and did not develop symptoms, and more than 7 days have passed since the food employee was diagnosed.

Managing Exclusions and Restrictions	2-201.13 Removal, Adjustment, or Retention of Exclusions and Restrictions (continued)
Sore throat with fever— removing exclusion or restriction	(G) Reinstate a food employee who was excluded or restricted as specified under subparagraphs 2-201.12(G)(1) or (2) if the food employee provides to the person in charge written medical documentation from a health practitioner stating the food employee meets one of the following conditions: (1) Has received antibiotic therapy for *Streptococcus pyogenes* infection for more than 24 hours, (2) Has at least one negative throat specimen culture for *Streptococcus pyogenes* infection, or; (3) Is otherwise determined by a health practitioner to be free of a *Streptococcus pyogenes* infection.
Uncovered infected wound or pustular boil—removing restriction	(H) Reinstate a food employee who was restricted as specified under ¶ 2-201.12(H) if the skin, infected wound, cut, or pustular boil is properly covered with one of the following: (1) An impermeable cover such as a finger cot or stall and a single-use glove over the impermeable cover if the infected wound or pustular boil is on the hand, finger, or wrist, (2) An impermeable cover on the arm if the infected wound or pustular boil is on the arm, or; (3) A dry, durable, tight-fitting bandage if the infected wound or pustular boil is on another part of the body.
Exposure to foodborne pathogen and works in food establishment serving HSP— removing restriction	(I) Reinstate a food employee who was restricted as specified under ¶ 2-201.12(I) and was exposed to one of the following pathogens as specified under subparagraph 2-201.11(A)(4) or (5):
Norovirus	(1) Norovirus and one of the following conditions is met: (a) More than 48 hours have passed since the last day the food employee was potentially exposed, or; (b) More than 48 hours have passed since the food employee's household contact became asymptomatic.
***Shigella* spp., EHEC, or STEC**	(2) *Shigella* spp. or Enterohemorrhagic or Shiga toxin-producing *Escherichia coli* and one of the following conditions is met: (a) More than 3 calendar days have passed since the last day the food employee was potentially exposed, or; (b) More than 3 calendar days have passed since the food employee's household contact became asymptomatic.
S. Typhi	(3) *S.* Typhi and one of the following conditions is met: (a) More than 14 calendar days have passed since the last day the food employee was potentially exposed, or; (b) More than 14 calendar days have passed since the food employee's household contact became asymptomatic.

(cont.)

Managing Exclusions and Restrictions	2-201.13 Removal, Adjustment, or Retention of Exclusions and Restrictions *(continued)*

Hepatitis A virus	(4) Hepatitis A virus and one of the following conditions is met:
	
	 (a) The food employee is immune to Hepatitis A virus infection because of a prior illness from Hepatitis A virus,;
	 (b) The food employee is immune to Hepatitis A virus infection because of vaccination against Hepatitis A virus,
	 (c) The food employee is immune to Hepatitis A virus infection because of IgG administration,
	 (d) More than 30 calendar days have passed since the last day the food employee was potentially exposed,
	 (e) More than 30 calendar days have passed since the food employee's household contact became jaundiced, or;
	 (f) The food employee does not use an alternative procedure that allows bare hand contact with ready-to-eat food until at least 30 days after the potential exposure, as specified in subparagraphs (I)(4)(d) and (e) of this section, and the food employee receives additional training about:
	    (i) Hepatitis A virus symptoms and preventing the transmission of infection,
	    (ii) Proper handwashing procedures, and;
	    (iii) Protecting ready-to-eat food from contamination introduced by bare hand contact.

Appendix E

Conversion Table for Fahrenheit and Celsius for Common Temperatures Used in Food Establishments

°F	°C	°F	°C
212	100	85	30
200	93	75	24
194	90	70	21
190	88	68	20
180	82	55	13
171	77	50	10
165	74	45	7
160	71	41	5
155	68	38	3
150	66	36	2
145	63	34	1
140	60	32	0
135	57	30	-1
130	54	28	-2
120	49	0	-18
110	43	-4	-20
100	38	-31	-35

Appendix F

Areas of Knowledge Deemed Important for the Person in Charge

Paragraph 2-102.11 (C) of the current *Food Code* identifies the following areas of knowledge as being important for the person in charge of a food establishment:

1. Describing the relationship between the prevention of foodborne disease and the personal hygiene of a food employee.

2. Explaining the responsibility of the person in charge for preventing the transmission of foodborne disease by a food employee who has a disease or medical condition that may cause foodborne disease.

3. Describing the symptoms associated with the diseases that are transmissible through food.

4. Explaining the significance of the relationship between maintaining the time and temperature of PHF (TCS) and the prevention of foodborne illness.

5. Explaining the hazards involved in the consumption of raw or undercooked meat, poultry, eggs, and fish.

6. Stating the required food temperatures and times for safe cooking of PHF (TCS) including meat, poultry, eggs, and fish.

7. Stating the required temperatures and times for the safe refrigerated storage, hot-holding, cooling, and reheating of PHF (TCS).

8. Describing the relationship between the prevention of foodborne illness and the management and control of the following:
    a. Cross contamination
    b. Hand contact with ready-to-eat foods
    c. Hand washing
    d. Maintaining the food establishment in a clean condition and in good repair.

9. Describing foods identified as major food allergens and the symptoms that a major food allergen could cause in a sensitive individual who has an allergic reaction, and assuring that employees are properly trained in food safety, including food allergy awareness, as it relates to their assigned duties.

10. Explaining the relationship between food safety and providing equipment that is:
    a. Sufficient in number and capacity, and;
    b. Properly designed, constructed, located, installed, operated, maintained, and cleaned.

11. Explaining correct procedures for cleaning and sanitizing utensils and food-contact surfaces of equipment.

12. Identifying the source of water used and measures taken to assure that it remains protected from contamination such as providing protection from backflow and precluding the creation of cross connections.

13. Identifying poisonous or toxic materials in the food establishment and the procedures necessary to assure that they are safely stored, dispensed, used, and disposed of according to law.

14. Identifying critical control points in the operation from purchasing through sale or service that when not controlled may contribute to the transmission of foodborne illness and explaining steps taken to assure that the points are controlled in accordance with the requirements of this *Code*.

15. Explaining the details of how the person in charge and food employees comply with the HACCP plan if a plan is required by the law, this *Code*, or an agreement between the regulatory authority and the food establishment.

16. Explaining the responsibilities, rights, and authorities assigned by this *Code* to the:

    a. Food employee
    b. Conditional employee
    c. Person in charge
    d. Regulatory authority.

17. Explaining how the person in charge, food employees, and conditional employees comply with reporting responsibilities and exclusion or restriction of food employees.

Appendix G

New Risk Designations for *Food Code* Provisions

The FDA has created a new three-tier system to designate *Food Code* provisions in terms of their relationship to the risk factors most likely to contribute to foodborne illness and the public health interventions and good retail practices that result in safer food and protect the consumer. "Priority," "Priority Foundation," and "Core" were the terms chosen to describe the relationship of the *Food Code* provisions to preventing, eliminating or reducing to an acceptable level, hazards associated with foodborne illness. The definitions for these three risk designations are as follows:

◆ **Priority** - means a provision in the *Food Code* whose application contributes directly to the elimination, prevention or reduction to an acceptable level of hazards associated with foodborne illness or injury, and there isn't another provision that more directly controls the hazard.

Priority items include those provisions in the *Food Code* which have a quantifiable measure or critical limit to show control of the hazard such as cooking, reheating, cooling, and hand washing. A priority item is denoted in the *Code* with the superscript letter ^{P}. When considering if a provision in the *Code* is a priority item, you should ask "Is there a measurable critical limit?" and "Do other provisions in the *Code* more directly control the hazard?"

◆ **Priority foundation** - means a provision in the *Code* whose application supports, facilitates, or enables the active managerial control of one or more priority items.

Priority foundation includes the purposeful incorporation of specific actions, equipment or procedures by industry management to attain control of factors that contribute to foodborne illness or injury such as personnel training, infrastructure or necessary equipment, HACCP plans, documentation or record keeping, and labeling. Priority foundation item is denoted in the *Food Code* with the superscript letters ^{PF}.

◆ **Core** - means a provision in the *Food Code* that is not designated as a priority item or a priority foundation item. Core items include an item that usually relates to general sanitation, operational controls, sanitation standard operating procedures (SSOPs), facilities or structures, equipment design, or general maintenance.

Appendix H

Factors Affecting Cleaning and Sanitizing

Factors that Affect Cleaning Efficiency		
Factor	**Impact on Cleaning**	**Description**
Type of and amounts of soil to be removed	Type and amount of soil determines cleaning agents and process used to remove soil	Soil consists of: ◆ Food deposits (proteins, carbohydrates, fats and oils) ◆ Mineral deposits (salts) ◆ Microorganisms (bacteria, viruses, yeasts, and molds) ◆ Dirt and debris.
Water quality	Affects efficiency of cleaning agent	◆ Must be potable (safe to drink) water ◆ Cleaning agents must be compatible with the characteristics of your water supply.
Detergent or cleaner to be used	Cleaning agent or solvent—dissolves and loosens soil	◆ See cleaning chart for more detailed information on page 288.
Concentration of cleaner	Using the recommended amount of detergent improves cleaning power	◆ Using too much detergent is a waste of money and may not improve cleaning.
Amount of time detergent/cleaner remains in contact with the surface	Reduces scrubbing necessary to remove soil	◆ Soaking items increases cleaning efficiency.
Water temperature	Increased water temperature helps decrease the strength of the bonds that hold soil to the surface	◆ Hot enough to remove soil but not to bake it on ◆ Heat-stable detergents work best when the water temperature is 130°F (54°C) - 160°F (71°C).
Velocity or force	Removes soil and film from food-contact surfaces	◆ Scrubbing or force to move soil/film from food-contact surfaces ◆ Less force is required when detergent is working correctly.

Different Types of Detergents and Cleaners

Cleaning Agent	Advantages	Disadvantages
Soaps	◆ Effective for hand washing in soft water ◆ Limited applications as cleaners	◆ Form precipitates and films in hard water ◆ Not compatible with some sanitizers.
Alkaline detergents	◆ Good general purpose cleaners ◆ Dissolve proteins and other organic material	◆ Strong alkalis are corrosive and can harm metals, equipment surfaces, and skin.
Acid detergents	◆ Frequently used to remove food, mineral deposits, and hard water deposits from the surfaces of equipment and utensils	◆ Strong acids are corrosive to metals and irritating to skin.
Degreasers	◆ Remove grease and oily soils from hard surfaces ◆ May be used for pre-treatment	◆ Can be irritating to skin and can leave a residue.
Abrasives	◆ When mixed with a detergent are useful for jobs that require scrubbing, scouring, or polishing	◆ Can scratch equipment surfaces ◆ Abrasive particles may also contaminate food.
Detergent sanitizers	◆ Effectively clean and sanitize a food-contact surface when applied to a food-contact surface 2 times—to clean the surface and to sanitize it ◆ Product may be used at 2-bay sink where there is no distinct water rinse between the washing and sanitizing steps	◆ Can leave chemical residue if used at too high a level of concentration.

Commonly Used Sanitizers

Sanitizing Agent	Advantages	Disadvantages
Chlorine compounds ◆ Commonly used **chlorine** compounds include liquid chlorine, hypochlorites, inorganic chloramines and organic chloramines ◆ The germicidal effectiveness of chlorine-based sanitizers depends, in part, on water temperature and the pH of the sanitizing solution	◆ Kills many types of microbes ◆ Good for most sanitizing applications ◆ Deodorize and sanitize ◆ Nontoxic to humans when used at recommended concentrations ◆ Colorless and nonstaining ◆ Easy to handle ◆ Economical to use	◆ Corrosive to equipment ◆ Can irritate human skin and hands.
Iodophors ◆ The iodine-containing sanitizers commonly used in food establishments are called **iodophors** ◆ They function best in water that is acidic and at temperatures between 68°F(20°C) and 120°F (49°C) ◆ Iodophors must be applied at 12.5 parts per million (ppm) when immersion sanitizing and at 25 parts per million in swab and spray applications	◆ Less corrosive to equipment ◆ Less irritating to skin ◆ Effective against a wide range of bacteria, small viruses and fungi ◆ Especially good for killing microbes on hands ◆ Iodophors kill more quickly than either chlorine or the quaternary ammonium compounds	◆ Moderate cost ◆ Can stain equipment.
Quats ◆ **Quaternary ammonium compounds (quats)** are ammonia salts used as chemical sanitizers in food establishments	◆ Stable at high temperature ◆ Stable for a longer contact time ◆ Good for in-place sanitizers ◆ Noncorrosive ◆ No taste or odor	◆ Very expensive ◆ Hard water can reduce effectiveness ◆ Destroy a narrower range of microorganisms compared to chlorine or iodophors, which may limit their use in some food establishments ◆ At concentrations above 200 ppm, quats can leave an undesirable residue on the surface of an item.

Factors that Affect The Action of Chemical Sanitizers

Factor	Description
Contact of sanitizer	◆ In order for the chemical to react with and destroy microorganisms, it must achieve close contact.
Selectivity of sanitizer	◆ Some sanitizers are nonselective and destroy a wide variety of microorganisms. Others exhibit a certain degree of selectivity.
Concentration of sanitizer	◆ In general, increasing the concentration of a chemical sanitizer proportionately increases its rate of microbial destruction. But there are limitations as the increased activity only extends to a certain maximum concentration, and then levels off. ◆ More is not always better, and high concentrations of sanitizers can be toxic and wasteful. Always follow the manufacturer's label use instructions to assure peak effectiveness of chemical sanitizers.
Temperature of solution	◆ All the common sanitizers increase in activity as the solution temperature increases. ◆ The standard range of water temperatures for chemical sanitizing solutions is between 68°F (20°C) and 120°F (49°C). ◆ Water temperatures as low as 55°F (13°C) can be used with chlorine under special circumstances, and temperatures above 120°F (49°C) should be avoided when using chlorine and iodine. At high temperatures, the potency of these sanitizers is lost by its evaporation into the atmosphere.
pH or acidity of solution	◆ Water hardness can affect the pH of water that exerts a significant influence on most sanitizers. ◆ Most soaps and detergents are alkaline with a pH between 10 and 12. That is why soap and detergent must be rinsed off the surfaces of equipment and utensils before they are sanitized.
Time of exposure	◆ Allow sufficient time for chemical reactions to destroy the microorganism. The amount of exposure time depends on the preceding factors as well as the size of the microbial populations and their susceptibility to the sanitizer.

Index

A

Abrasive cleaning, 156
Accident prevention, 11
Acidity, 25
Acquired Immune Deficiency Syndrome (AIDS), 77
Added man-made chemicals, 53–54
Adulterated, 96
Aerobic bacteria, 26
AIDS, 77
Air gap, 186
Alkaline, 25
Allergens, 22, 49–50, 245, 251, 253, 267, 283
    labelling, 234–235
Allergies. *See* Allergens
Ambient temperature, 157
American Gas Association (AGA), 130
American National Standards Institute (ANSI), 130
Anaerobic bacteria, 26
Anisakis spp., 45, 245, 267
Answers to end of chapter questions, 261
Anti-microbial rinse, 103
Anti-slip mats, 182
Approved sources, 89
Atmospheric vacuum breaker, 186
At-risk population. *See* Immunocompromised
At-risk populations, 7
A_w. *See* Water activity

B

Bacillus cereus, 33–34, 71, 245
    illness statistics, 63
    sources of, 208
Backflow, 185
    carbonators, 187
    preventing, 185
Backpressure, 185
Backsiphonage, 185
Bacteria, 22–23, 30
    acidity, 25
    defined, 22
    food, 25
    foodborne illnesses, 30–42
    growth of, 23

    moisture, 26
    multiplying, 24
    non-sporeforming, 36–38, 40–42
    oxygen, 26
    pathogenic, 23
    spoilage, 23
    sporeforming, 31
    temperature, 25
    time, 26
Bacterial growth, 23
Bait box, 193
Band saw, 166
Be Food Safe, 238
Beetles, 191
Big Five pathogens
    reporting employee illness, 271–279
Bi-metal thermometer, 65
Binary fission, 23
Biological hazards, 30, 32–49
    defined, 22
Bioterrorism, 219
Boil water advisories, 217
Boil-point method, 67
Bottled drinking water, 217
Bucket and brush method, 143
Building condition, 181
Bulging cans, 93
Butter, 100

C

Calibrating a thermometer, 66
    boiling point method, 67
    ice point method, 67
Campylobacter jejuni, 36, 96, 247
CAP. *See* Controlled atmosphere packaging, 92
Caps and hair or beard restraints, 76
Carbonators, 187
Carrier, 247
CCPs. *See* Critical control points
CDC *See* Centers for Disease Control and Prevention
Ceilings, 169, 183
Celsius-fahrenheit conversion table, 281
Centers for Disease Control and Prevention, 4–5, 20, 55, 62–63, 89, 122, 232

Certification, food manager, 12
Certified food protection manager, 13
CFP. *See* Conference for Food Protection
CFR. *See* Code of Federal Regulations
Cheese, 100
Chemical hazards, 22, 49–54
    added man-made chemicals, 53–54
    employee medication, 54
    food allergens, 50
Chemical pesticides, 189
Chemical sanitizing, 161
Chemical sanitizing solutions, 162
Chemical storage, 107
Chemical test strips, 162, 166
Chemicals
    foodborne illness caused by, 49–54
Children's menu, 114
Chlorinating water, 217
Chlorine, 161
Choppers, 140
Ciguatoxin, 51
    illness statistics, 63
Clean clothing, 80
Cleaning agent, 155
Cleaning and sanitizing, 142, 154
    abrasive cleaning, 156
    ceilings, 169
    chemical sanitizing, 161
    chlorine, 161
    cleaning agents, 155
    cleaning, defined, 10
    clean-in-place systems, 156
    detergents and cleaners, 156
    environmental areas, 169–170
    equipment and supplies, 170
    five steps of, 155
    fixed equipment, 166
    floors, 169
    food-contact surfaces, 159
    frequency, 157
    heat sanitizing, 159, 161
    hot water, 160
    in-place equipment, 168
    manual warewashing, 165–166
    mechanical warewashing, 163–164
    pre-flushing, 155
    removal of food particles, 155
    rinsing, 157
    sanitary, defined, 10
    soaking, 156

    spray methods, 156
    steam, 160
    walls, 169
    wiping cloths, 168
Clean-in-place systems, 156, 168
Clostridium botulinum, 35, 71, 93–94
Clostridium perfringens, 34, 62–63, 71, 247, 267
    illness statistics, 63
    sources of, 208
CLs. *See* Critical limits
Cockroaches, 190
Code of Federal Regulations (CFR), 79
Coding, 216
Codworm, 45
Cold food
    recommendations for, 71–72
Cold storage, 106, 111–112
Cold-holding, 111–112
Commingling, 102
Conference for Food Protection (CFP), 12, 232
Construction materials, 132
Consumer advisory, 237–238
Contains statement (allergens), 237
Contamination, 9–12, 73
    defined, 8
    sources of, 8, 79
Controlled atmosphere packaging (CAP), 92
Controlling temperatures, 64, 68
Convection oven, 135
Conversion table (fahrenheit-celsius), 281
Cook-chill, 92
    refrigeration, 139
Cooking, 71, 113–115
    foods recommendations for, 68
    temperatures/times, 71
Cooling, 115–116
    foods-danger of, 71–72
    foods-guidelines for, 68
    methods, 116
Core item, 230
Counter-mounted equipment, 132
Coving, 182
Crisis management, 11, 217–221
Critical control points (CCPs), 208–209
    monitoring, 212
Critical limits (CLs), 210–213
Cross connection, 185
Cross contamination, 78–79, 106, 110, 120, 202, 208
    preventing, 78

Cryptosporidium parvum, 48
Cyclospora cayetanensis, 46, 103, 249, 267

D

Date marking, 235–237
Deck oven, 135
Definitions (glossary), 245–260
Degreasers, 156, 288
Dented cans, 93
Design, layout, and facilities, 130
Detergents, 156
Dial-faced, metal stem type thermometer, 65
Digital thermometer, 65
Discarding food, 118
Disclosure (consumer advisory), 237
Disease-causing bacteria. *See* pathogenic
    bacteria
Dishwashing
    See warewashing equipment, 142
Disposable gloves, 75
    using, 75
Dry milk, 99
Dry storage, 107
Dual check valve, 186

E

E. coli, 6–7, 37, 77, 233, 250, 257, 268, 271–277
    Big Five pathogen, 271
    case study, 240
    fruits and vegetables, 103
    illness statistics, 63
    in- ground meat, 208
    reporting illness, 77
    reporting, restrictions and exclusions, 271–
        279
Eating and smoking, 76
Eggs
    egg products, 99
    receiving, 98–99
Emergency procedures, 217–221
Employee health
    reportable illnesses, 271–279
End-of-chapter quizzes, answers, 261
Environmental Protection Agency (EPA), 232
Environmental sanitation and maintenance,
    182–195
    ceilings, 183
    exterior of building, 181
    floors, 182
    garbage, 187
    handwashing facilities, 184
    pest control, 189, 194
    plumbing hazards, 185–187
    restrooms, 183
    sewage disposal, 180
    walls, 183
    water supply, 185–187
EPA, 232
Equipment, 130
    choppers, 140
    cleaning fixed, 166
    cleaning in-place equipment, 168
    construction materials, 132
    cook-chill, 92
    grinders, 140
    hot-holding, 139
    ice machines, 141
    installation, 144
    low temperature storage, 136
    maintenance, 144
    mixers, 140
    other types of, 140–141
    ovens, 135
    rapid-chill, 139
    refrigeration, 136–139
    replacement, 144
    selection, 131, 134
    slicers, 140
    types of, 134–136, 143
    types of material, 133
    warewashing, 142
        manual, 143
        mechanical, 143
Escherichia coli. See E. coli
Exterior of building, 180

F

FAAN. *See* Food Allergy and Anaphylaxis
    Network
Facilities, 130
Facilities *See* Equipment
Facultative anaerobic, 26
Fahrenheit-celsius conversion table, 281
FALCPA. *See* Food Allergen Labeling and
    Consumer Protection Act, 234
FATTOM, 24, 26–27
FDA *Food Code*, 12, 231
FDA. *See* Food and Drug Administration

Federal agencies, 231–232
Federal Trade Commission (FTC), 232
FIFO
    First-in, first-out, 104
Fight BAC!, 238
Fish, 100–102
    receiving, 100
Fixed equipment
    cleaning, 166
Flies, 190
Floor drains, 169
Floor-mounted equipment, 132
Floors, 169, 182
Fluid milk, 99–100
Food
    allergens, 267
    discarding, 118
    frozen, ready-to-eat, 112
    mishandled, 63
    prepackaged, 113
    reconditioning, 118
    serving, 118
    temperature recommendations, 68
    temperatures, measurement of, 64–66
Food additives, 22, 49, 53, 144, 231
Food Allergen Labeling and Consumer
    Protection Act (FALCPA), 234–235
Food Allergy and Anaphylaxis Network
    (FAAN), 50
Food and Drug Administration (FDA), 231
Food establishment
    condition of, 180
    defined, 4
Food labeling, 233, 235–238
    allergens, 234
Food manager certification, 12
Food product flow, 88
    cold-holding, 111–112
    cooking, 113–115
    cooling, 115–116
    discarding food, 118
    handling, 118
    hot-holding, 117
    inspecting delivery vehicles, 89
    packaging, 91–95
    preparation, 110
    receiving, 89–104
    reconditioning, 118
    reheating, 117
    serving, 118
    storage, 104–109
    thawing, 110–111
Food recall, 216–217
Food safety
    organizations, 232
    recent initiatives, 238
Food safety management, 202, 220
    programs, 202, 215
Foodborne disease outbreak
    defined, 7
Foodborne hazard. See Hazards
Foodborne illness, 218
    causes, 7
    classifications of, 21
    defined, 4
    factors, 62–63
    incident, 218
    symptoms, 20
Foodborne illness incident
    action plan, 219
Foodborne infection, 20, 22, 36
Foodborne parasites, 45–48
Food-contact surfaces, 132
Four Rs (food allergy), 50
Freezer storage, 106
Frozen foods, 71, 113
    receiving, 104
Fruits and vegetables, 103
Fryer, 135
FTC, 232
Fungi, 48

G

Game animals, 97
GAPs. See Good agricultural practices
Garbage, 187
Generally recognized as safe (GRAS), 233
Germs. See Microorganisms
Glossary, 245–260
Gloves
    proper use of, 75
Glueboard, 193
GMPs. See Good manufacturing practices
Good agricultural practices (GAPs), 215
Good health, 80
Good manufacturing practices (GMPs), 215
Good retail practices (GRPs), 215
Government regulations. See Food safety
    regulations

Grade standards, 95
Grading, 233
Grease traps, 187
Grinders, 140
GRPs. *See* Good retail practices

H

HACCP system, 10–11, 203–215
    development of food flow diagram, 203–205
    hazard analysis, 206
    seven principles of, 206–215
Hair restraints, 80
Hand antiseptics, 75
Hand washing, 184
    facilities, 184
    good personal hygiene, 73
    proper technique for, 73–75
Hazard Analysis Critical Control Point
    System. *See* HACCP
Hazards, 22
    biological hazards, 22
    chemical hazards, 49–54
    physical hazards, 22
Health
    personal, 77
Heating, ventilation, and air conditioning
    (HVAC), 146
Heat-shocked, 32
Hepatitis A virus, 6, 43–44, 268, 271–276, 279
    case study, 172
    fruits and vegetables, 103
    good hygiene, 73
    raw seafood, 101
    reporting illness, 77
    reporting, restrictions and exclusions, 271–
        279
Hermetic packages, 92
Highly susceptible population. *See*
    Immunocompromised.
Histamine poisoning, 52
HIV virus, 77
Holding thermometer, 161
Home meal replacement, 121
Home-canned foods, 35
Hood systems, 146
Hoof- (foot-) and-mouth, 55
Hot food, recommendations for, 71
Hot-holding, 116–117
    equipment, 139
House mouse, 191

HVAC - heating, ventilation, and air
    conditioning, 146
Hygiene practices, 80

I

Ice machines, 141
Illnesses
    reportable, 77, 271–279
Immunocompromised, 7, 21
Improper cooling of foods, 72
Infection. *See* Foodborne infection
Infrared thermometer, 65
Ingredient substitution, 110
Insects, 189–191
Inspection, 229–230
Inspection for wholesomeness, 233
    grading of food, 233
Inspection provisions, 230
Installation of equipment, 144
Integrated pest management (IPM), 189, 194
Intoxication, 20–22, 33, 35, 41–42, 51, 253
Iodophors, 161
IPM. *See* Integrated pest management
Irradiation, 95

J

Jewelry, 76
Juice, 104

K

Kitchenware, 142
Knowledge requirements, person in charge,
    272, 274

L

Lighting, 144–145
Listeria monocytogenes, 6–7, 38, 94, 111, 233,
    236, 253, 268
    growth temperature, 25

M

Mad cow disease, 55
Manual warewashing, 165–166
MAP. *See* Modified atmosphere packaging
Material safety data sheet (MSDS), 170–171
Materials, 133
Maximum registering (holding)
    thermometer, 66

Measuring food temperature, 67
Meat/meat products, 95–96
Mechanical warewashing, 163–164
Medication, 54
Metals, 133
Microbes, *See* Microorganisms
Microorganisms, 9
Microwaves, 136
Milk products, 99
Mobile food facilities, 119
Modified atmosphere packaging (MAP), 92–93
Moths, 191
MSDS. *See* Material safety data sheet
Mycotoxin, 48

N

National Marine Fisheries Service
    (NMFS), 231
Non-food-contact surfaces, 132
Nonpotable water, 181
Norovirus, 43, 62, 268, 272–275, 277–278
    Big Five pathogen, 271
    case study, 20, 55
    fruits and vegetables, 103
    illness statistics, 63
    reporting illness, 77
    reporting, restrictions and exclusions,
        271–279
Norway rat, 191
NSF International, 130

O

Occupational Safety and Health
    Administration (OSHA), 232
Onset time, 21
Organizations, 232–233
OSHA. *See* Occupational Safety and Health
    Administration
Outbreak, 218
    action plan, 219
Ovens, 135–136

P

Packaged foods, 91–95
Parasites, 45–48
    foodborne illness caused by, 45–48
Pasteurization, 99
Pathogenic bacteria, 23

Permit to operate, 228
Person in charge, 229, 235
    knowledge requirements, 272, 274
Personal habits, 76
Personal hygiene, 6, 9, 73, 76, 78–79, 205,
    229, 255
    clean uniforms, 75–76
    clothing and apparel, 75–76
    disposable gloves, 75
    GMPs, 215
    hair restraints, 76
    recommendations for, 73, 77
    reportable diseases, 77
    toilet facilities, 183
Pest control, 189–194
Pests, 79
pH, 25
PHF (TCS), 28–29, 255
    cleaning guidelines for food-contact
        surfaces, 159
    cooking guidelines, 114
    cooling, 115
    date marking, 235–236
    defined, 28
    determining, 263–265
    food-contact surfaces, 157
    foods, 103
    hazard analysis, 206
    holding foods, 89
    holding temperatures, 71
    hot-holding, 117
    non-PHF (TCS) products, 29, 37, 91, 118
    ready-to-eat foods, 117
    temperature control, 68
    viruses, 43
Physical hazards, 22, 54
Plastics, 134
Plumbing hazards, 185–187
Post-inspection conference, 229
Potable water, 154
Potentially hazardous food (time/temperature
    control for safety food). *See* PHF (TCS)
Poultry
    receiving, 96–97
Pre-flushing, 155
Pre-inspection conference, 229
Prepackaged foods, 113
Preparation and service, 110–113
Preserving food, 92–94
Pressure type vacuum breaker, 186

Priority item, 230
Product coding, 216
Product dating labels, 235
Product tampering, 220
Professional organizations, 232–233
Purchasing, 89
    approved sources, 89

Q

Quaternary ammonium compounds (quats),
    161, 289
Quizzes, answers, 261

R

Radura, 95
Rapid-chill refrigeration, 139
Rats. *See* Rodents
Reach-in refrigeration, 138
Ready-to-eat foods, 30, 121, 256
    cross contamination, 78–79
    date marking, 235–237
    defined, 30
    examples of, 30
    frozen, 113
    holding, 112, 117
    hygiene, 76
    inspecting, 233
    prepackaged, 113
    storage, 78, 106
    storage of PHF (TCS), 117
    temperature, measuring, 66
    thawing, 110–111
    touching, 75
    vending machines, 120
Recall of foods. *See* Food recall
Receiving, 89–104
    food quality, 90
    packaged foods, 91
    temperatures, 89
Reconditioning food, 118
Red meat products, 95–96
Reduced oxygen packaged foods (ROP), 35
Reduced oxygen packaging, 92–95
Reduced pressure principle backflow
    preventers, 186
Refrigerated foods, 71
Refrigerated storage, 105–106
Refrigeration
    display, 138

    equipment, 136
    reach-in, 138
    thermometers, 66
Refrigeration equipment, 137–139
Refuse, 187
Regulations
    local, 228
Regulatory considerations, 130
Reheating, 71, 117
Removing food particles, 155
Restrooms, 183
Returnable containers
    Refilling returnable containers, 118
Rinsing, 157
Rodenticides, 193
Rodents, 79, 191, 194
    controlling, 193–194
    signs of infestation, 191–193
Roof rat, 191
ROP. *See* Reduced oxygen packaging
Roundworms, 45, 47
RTE. *See* Ready-to-eat foods
Rusty cans, 93

S

Safe food temperatures, 65, 89
Safe food-handling label, 234
Safe load line, 113
Salmonella spp., 4, 6–7, 14, 39, 62–63, 77,
    257, 268
    and hygiene, 62
    case study, 122, 202, 221
    eggs, 98–99, 108
    fruits and vegetables, 103
    illness statistics, 63
    inspection for, 233
    poultry, 96
    raw meat, 208
    reporting illness, 77
    reporting, restrictions and exclusions,
        271–279
Sanitary, 79
    defined, 10
Sanitizing. *See* Cleaning and sanitizing
Scombrotoxin, 52
    illness statistics, 63
Seafood, 100–102, 109
Self-inspections, 229
Self-service bar, 119

Sell-by date, 94, 216
Sensing portion of thermometer, 67
Sensory evaluation, 90
Sewage disposal system, 180
Shelf life, 92
Shellfish, 101
    identification tags, 102
    receiving, 101
Shellfish tags, 102
Shiga toxin-producing *Escherichia coli.*
    See E. coli
Shigella spp., 40, 257, 268, 271–275, 277–278
    fruits and vegetables, 103
    hygiene, 73
    illness statistics, 63
    reporting illness, 77
    reporting, restrictions and exclusions,
        271–279
Shigellosis, 40, 272
Single-service articles, 141–142
Single-use articles, 141–142
Slicers, 140
Soaking, 156
SOPs. *See* Standard operating procedures
Sous vide, 92, 94
Spoilage bacteria, 23
Spoilage in food, 90
Sporeforming bacteria, 31–35
Spores, 22, 31
Spray cleaning methods, 156
Stainless steel, 133
Standard operating procedures (SOPs), 209,
    212
Staphylococcus aureus, 41, 62, 73, 77, 80, 208,
    269
    illness statistics, 63
State and local regulations, 228
Steam cleaning, 160
Stock rotation, 105
Storage, 104–109
    chemical, 107
    condition of foods, 107–109
    dry, 107
    freezer, 106
    frozen, 106
    inside refuse, 188
    outside refuse, 188
    refrigerated, 105–106
    types of, 105

T
Tableware, 142
Tasting, 79
Temperature abuse, 64
    avoiding, 110
    preventing, 68, 71–72
Temperature and time abuse. *See*
    Temperature abuse
Temperature danger zone, 25, 29, 32, 64,
    71–72, 258
    cooling foods through, 115, 136, 139
    hot-holding, 117
    limiting time in, 110
    temperature and time abuse, 110
Temperature guidelines, 71
Temperature-measuring devices
    bi-metal thermometer, 65
    calibrating, 66
    calibration of, 66–67
    cleaning and sanitizing of, 66
    dial-faced, metal stem type thermometer,
        65
    digital, 65
    guidelines, 66
    infrared, 65
    maximum registering (holding), 66
    measuring temperatures, 67
    refrigerator/freezer, 66
    requirements of, 64
    thermocouple, 65
    T-Sticks, 65
    types of, 66
    using, 67
Temperatures
    holding, 71
    measuring, 65
Temporary facilities, 119
Test strips, 166
Thawing, 72, 110–111
Thermocouple, 65
Thermometers. *See* Temperature-measuring
    devices
Three-compartment sink, 143
Time and temperature abuse. *See*
    Temperature abuse
Toilet facilities, 183
Toxin-mediated infection, 20–22, 33–34, 37, 40
Toxins, 22, 49
Tracking powder pesticides, 194

Trash baler, 188
Tribal agencies, 228
Trichinella spiralis, 47, 259, 269
T-Sticks, 65

U

U.S. Department of Agriculture (USDA), 231
U.S. Department of Commerce (USDC), 231
UHT, 99
Underwriters Laboratories, Inc. (UL), 130
Uniforms, 75–76
Use-by date, 94, 216
Utensils, 142

V

Vacuum breaker, 186
Vacuum packaging, 92–93
Vegetables, 103
Vegetative cells, 22, 31
Vegetative state, 22, 32
Vending machines, 120
Ventilation system, 146
Vibrio spp., 42, 269
Viruses, 9, 22, 31, 43, 62, 259

W

Walk-in refrigeration, 138
Walls, 169, 183
Warewashing
    equipment, 142
    manual, 143, 165–166
    mechanical, 143
Warning label, 104
Water activity, 26–27, 29, 41, 209, 211, 260, 264
Water supply, 180
Water supply emergency, 217–218
Whole shell eggs, 98
Wholesome, 95–96, 203, 231, 233
Wiping cloths, 168
Wood, 134
Work centers, 131
Work clothes, 75